NCERT WORKBOOK BIOLOGY

Volume 2

CLASS XII

NCERT WORKBOOK
BIOLOGY

Volume 2
CLASS
XII

FOR SURE SHOT SUCCESS IN
NEET

Authors
Sanubia Saleem
Kavita Thareja
K. Anita

arihant
ARIHANT PRAKASHAN (Series), MEERUT

arihant

Administrative & Production Offices

Regd. Office

'Ramchhaya' 4577/15, Agarwal Road, Darya Ganj, New Delhi -110002
Tele: 011- 47630600, 43518550

Head Office

Kalindi, TP Nagar, Meerut (UP) - 250002, Tel: 0121-7156203, 7156204

Sales & Support Offices

Agra, Ahmedabad, Bengaluru, Bareilly, Chennai, Delhi, Guwahati, Hyderabad, Jaipur, Jhansi, Kolkata, Lucknow, Nagpur & Pune.

ISBN 978-93-25294-02-8

PO No : TXT-XX-XXXXXXX-X-XX

Published by Arihant Publications (India) Ltd.

For further information about the books published by Arihant, log on to www.arihantbooks.com or e-mail at info@arihantbooks.com

Follow us on

PREFACE

NCERT books are the most valuable assets for NEET aspirants. More than 90% questions asked in NEET 2019 and 2020 were based on concepts of NCERT. This simply means conquering NEET requires a firm grip over NCERT concepts. However, it is not so easy as NCERT is one of those theoretical books in which each line has many interpretations.

So far multiple resources are there in the market to help in gripping the NCERT concepts. But all of them are MCQs based practice books. Practice with all such resources makes the learning monotonous and boring.

"NCERT Workbook Biology" is altogether different from all such resources. This innovative book is written and developed keeping in mind the concept of 'Fun with learning'. The speciality of the book is, it makes you apply the conceptual knowledge in answering different types of questions which just require a word or maximum a sentence as their answer. In this way you can revise & strengthen your concept in an interesting manner.

A variable typology of questions given in each chapter, like;

- True-false
- Word correction in sentences
- Sequence arrangement and completion
- Small matchings
- Analogy completion
- Figure labelling and figure identification, etc.
- Fill in the blanks
- Sequential term arrangement
- Odd one out type
- Flow chart completion

Such a variety in question types reduces monotonicity of subject and increases concept understanding and figure visualisation.

This matchless fun filled practice book will help NEET aspirants in gripping the NCERT concepts to their maximum. We hope all of you will enjoy the learning along with practice while exploring this unmatched resource and secure sure shot success in NEET.

Authors

CONTENTS

1.	Reproduction in Organisms	1-10
2.	Sexual Reproduction in Flowering Plants	11-33
3.	Human Reproduction	34-52
4.	Reproductive Health	53-63
5.	Principles of Inheritance and Variation	64-84
6.	Molecular Basis of Inheritance	85-115
7.	Evolution	116-136
8.	Human Health and Disease	137-159
9.	Strategies for Enhancement in Food Production	160-177
10.	Microbes in Human Welfare	178-188
11.	Biotechnology : Principles and Processes	189-205
12.	Biotechnology and Its Applications	206-216
13.	Organisms and Populations	217-232
14.	Ecosystem	233-248
15.	Biodiversity and Conservation	249-262
16.	Environmental Issues	263-282

CHAPTER ONE

REPRODUCTION IN ORGANISMS

Topic 1 : Lifespan and Asexual Reproduction

1. The period from birth to the natural death of an organism represents its ____________ .
2. A mango tree has a ____________ (longer/shorter) lifespan as compared to a peepal tree.
3. Name an organism which is immortal.____________
4. Which characteristic of a living organism ensures continuity of species on earth?
   ____
5. Arrange the plants given below in their increasing order of lifespan.

Rose bush, Banana tree, Banyan tree, Rice plant

6. (i) Lifespan is a specific trait of each organism.

   True ☐ False ☐

   (ii) Lifespan is correlated with size and complexity of organisms.

   True ☐ False ☐

7. A biological process in which an organism gives rise to young ones similar to itself is ____________
8. Another name given to young ones of an organism. ____________
9. Give the three stages of the cycle of life. ____________
10. The organism's habitat, its internal physiology and several other factors are collectively responsible for how it reproduce.

    True ☐ False ☐

11. Match the following columns.

Column I (Organisms)	Column II (Lifespan)
A. Tortoise	1. 30 years
B. Horse	2. 60 years
C. Crocodile	3. 1-2 weeks
D. Butterfly	4. 100-150 years

A. ______________ B. ______________ C. ______________ D. ______________

12. Complete the following table.

Column I (Organisms)	Column II (Lifespan)
Elephant	1. (A)
(B)	2. 15 years
(C)	3. 140 years
Dog	4. (D)
Cow	5. (E)

A. ______________ B. ______________ C. ______________
D. ______________ E. ______________

13. Complete the following analogy with respect to their lifespan.

Mosquito : ______________ : : Fruitfly : ______________ .

14. Reproduction types in living organisms are based on participation of nucleus of an organism.

True ☐ False ☐

15. The offspring of asexual reproduction are ______________ to its parent.

16. Asexual reproduction is ______________ (slow/quick) and involves ______________ (one/two) parent(s).

17. The given process represents which type of asexual reproduction?

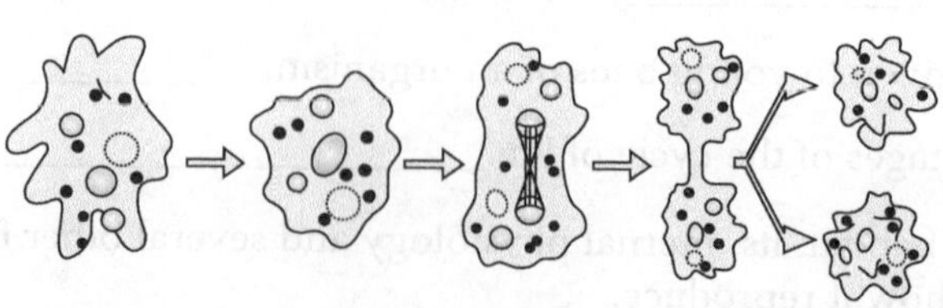

____

18. ______________ is a group of morphological and genetical identical individuals.

19. Name two organisms in which cell division is itself a mode of reproduction.

____

20. Asexual reproduction cannot involve _______________ (fusion/formation) of gametes.

21. The given vegetative propagule/asexual reproductive structures shown are found in.

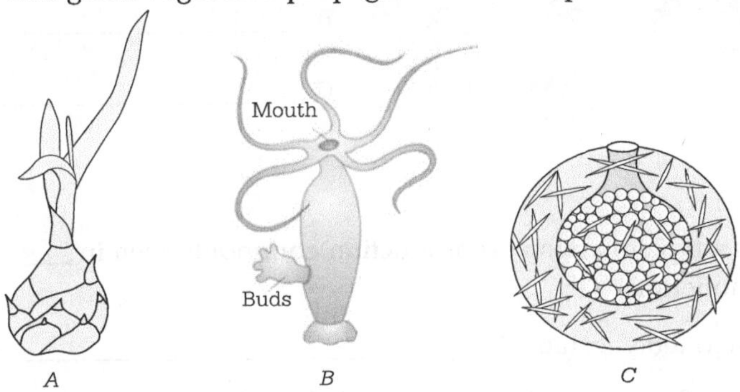

A. _______________

B. _______________

C. _______________

22. Complete the following analogy with respect to mode of reproduction.

Euglena : Binary fission : : Yeast : _______________

23. Select the odd one out with respect to reproductive structures.

Zoospores, Conidia, Buds, Gemmules, Oospores

____

24. Name a method of raising new plants from a small plant tissue with the help of tissue culture technique. _______________

25. Name the method of reproduction in single-celled organism in which their body divide into two/many parts and each part rapidly grows into an adult. _______________

26. _______________ is a phenomenon in which some single-celled organisms under unfavourable condition secrete a three-layered hard covering or cyst around itself.

27. In *Amoeba,* under unfavourable conditions, _______________ takes place but under favourable conditions, _______________ occurs.

28. Choose the mismatch pair.

(i) Zoospores – *Chlamydomonas* (ii) Buds – *Hydra*

(iii) Gemmules – *Paramecium* (iv) Conidia – *Penicillium*

____

29. Match the following columns.

	Column I (Asexual reproduction types)		Column II (Examples)
A.	Longitudinal binary fission	1.	*Planaria*
B.	Fragmentation	2.	*Vorticella*
C.	Spore	3.	*Ulothrix*

A. _______________ B. _______________ C. _______________

30. Examine the figure and write the asexual reproductive structure labelled parts as *A*, *B* and *C*.

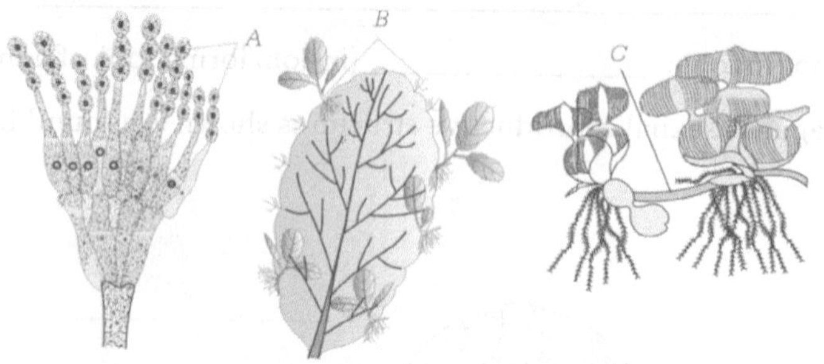

A. ____________________

B. ____________________

C. ____________________

31. Vegetative reproduction is a type of asexual reproduction commonly seen in ________ (plants/animals/protists/monerans).

32. Banana plant can be regenerated vegetatively by ____________________

33. What are the different units of vegetative propagation in plants? ____________________

34. A common term given to tuber, rhizome and bulb. ____________________

35. In some organisms, the body breaks into two pieces, each piece grows into an adult capable of producing offspring. This type of reproduction is called as <u>fission</u>.
(*Rewrite the statement by changing the underlined word.*)

____

36. Water hyacinth or Terror of Bengal is the most invasive weed found growing wherever there is stagnant water.

True ☐ False ☐

37. Vegetative reproduction is also a type ____________________ reproduction.

38. The site of origin of the new plantlets in potato, *Dahlia*, ginger and banana is <u>internodes of modified stem</u>.
(*Rewrite the statement by changing the underlined word.*)

____

39. In which two plants new plantlets arise from the same organ? ____________________

40. Match the following columns.

	Column I (Organisms)		Column II (Reproduce by)
A.	Ginger	1.	Tuber
B.	Garlic	2.	Sucker
C.	Potato	3.	Rhizome
D.	*Chrysanthemum*	4.	Bulb

A. __________ B. __________ C. __________ D. __________

41. Name any two groups of organisms which switch to sexual reproduction from asexual mode during adverse condition.

____

Topic 2 : Sexual Reproduction

42. When two parents (opposite sex) participate in the reproductive process and also involve fusion of male and female gametes it is called _______________

43. _______________ (Perennial/Annual/Seasonal) plants show clear-cut vegetative, reproductive and senescent phase.

44. The growth phase of an organism before attaining sexual maturity is referred to as

____

45. Rearrange the following phases in correct order.

Juvenile phase → Senescent phase → Reproductive phase → Pre-reproductive phase

____

46. Mark the incorrect match.
   (i) Neelakurinji – Perennial plant
   (ii) Bamboo (*Bambusa*) – Live vegetatively for 12 years
   (iii) Interflowering period – Mango, orange
   (iv) Biennial plants – Polycarpic

____

47. The plant named *Strobilanthes kunthiana* flowers once in _______________ years.

48. Name the phase, all organisms have to pass through before they can reproduce sexually.

____

49. The end of juvenile phase marks the beginning of senescent phase.

True ☐ False ☐

50. Complete the following analogy with respect to reproductive cycle of organisms.

Sheep : Oestrus cycle : : Apes : _______________

51. Oestrus cycle is cyclic change in the activities of ovaries and accessory duct in non-primates during_______________

52. Organisms reproducing throughout the year are called _______________ breeders, e.g. _______________ and those who show recurring sexual activity are called _______________ breeders, e.g. _______________ .

53. Identify the events (*A, B, D* and *E*) of sexual reproduction given below.

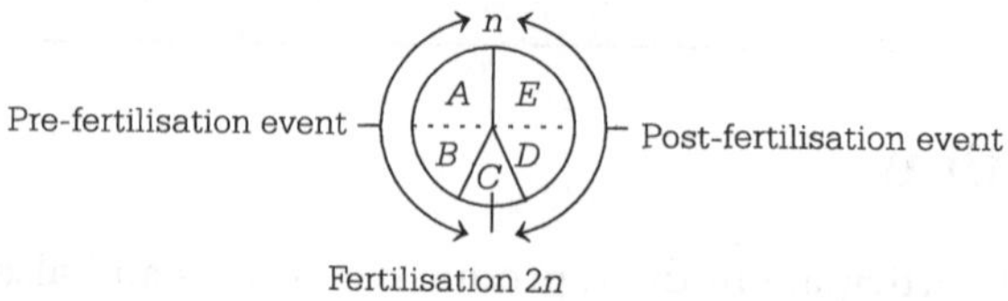

A. ____________________

B. ____________________

D. ____________________

E. ____________________

54. Name the process of formation of gametes. ____________________

55. Identify the type of gametes shown in figures *A*, *B* and *C*, respectively.

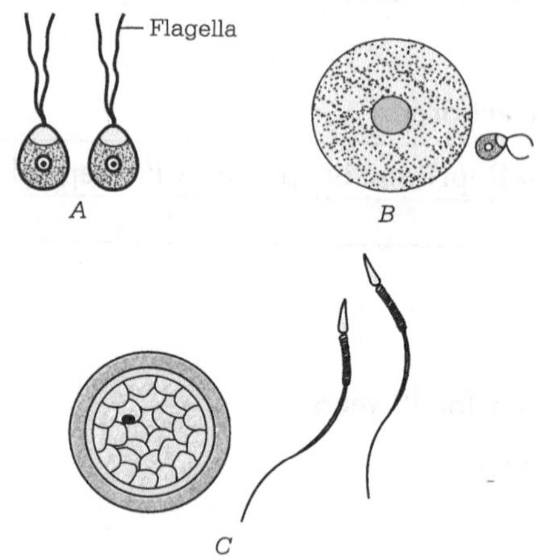

A. ____________________

B. ____________________

C. ____________________

56. Name the type of gametes found in algae. ____________________

57. What is male gamete called in heterogametic condition? ____________________

58. The condition, in which both male and female reproductive organs are found on the same plant is called unisexual.
(*Rewrite the statement by changing the underlined word.*)

____

59. In below given diagram of *Chara*, write down the names of the structures *A* and *B*.

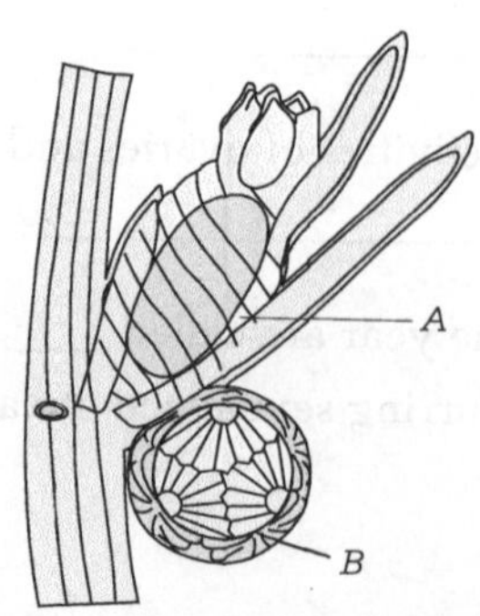

A. ____________________

B. ____________________

60. The condition in which male and female parts are present on different organisms, is called

____

61. In flowering plants, the unisexual male flower is called ____________, while the female is called ____________. Flowering plants may be monoecious, e.g. ____________ or dioecious, e.g. ____________.

62. The type of gametes formed in staminate and pistillate flowers are antherozoid and egg, respectively.

True ☐ False ☐

63. Which of the following is hermaphrodite (bisexual)?

Ant, Aphids, Earthworm, Cockroach

____

64. Which among the following is unisexual species?

Tapeworm, Leech, Fish

____

65. If the parent body is haploid then the gametes are ____________

66. In diploid organism, the gamete producing cells are called ____________

67. A haploid parent produces gametes by ____________.

68. Which of the following is an incorrect combination of organism with its chromosome number in meiocyte and in gamete?

Name of organisms	Chromosomes number in meiocyte	Chromosomes number in gamete
(i) Butterfly	380	190
(ii) Potato	16	08
(iii) Maize	20	10
(iv) Apple	34	17

____

69. Match the following columns.

Column I (Organisms)	Column II (Number of chromosomes)
A. Human	1. $2n = 12$
B. Fruitfly	2. $2n = 16$
C. Onion	3. $2n = 46$
D. Housefly	4. $2n = 08$

A. ________ B. ________ C. ________ D. ________

70. Identify the sequence of events shown in the diagram given below.

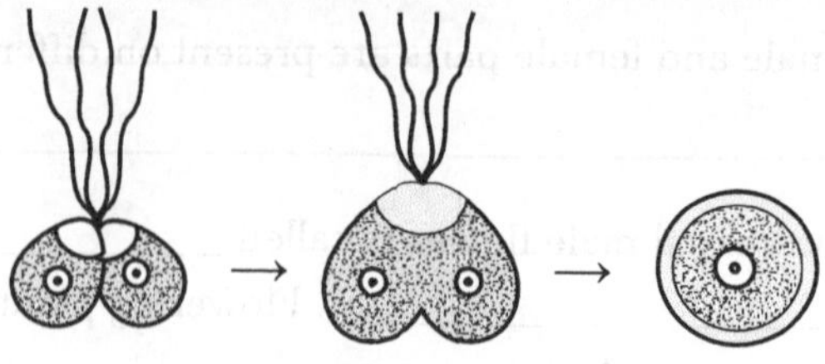

____

71. Self-fertilisation occurs in the <u>unisexual flower</u>.
(*Rewrite the statement by changing the underlined word.*)

____

72. In which of the following organisms self-fertilisation is seen?

Peas, Mustard, Sweet potato, Papaya

____

73. Essential and most critical event in sexual reproduction is fertilisation and formation of diploid zygote.

True ☐ False ☐

74. Fusion of male and female gametes is called heterogamy.

True ☐ False ☐

75. In some plants, the female gamete develops into embryo without fertilisation. This phenomenon is known as _______________

76. Syngamy may occur in _______________ medium.

77. The organisms exhibiting external fertilisation release a large number of _______________. These organisms show great _______________ between the sexes. In order to enhance the chances of _______________. To produce large number of _______________ as they are vulnerable to predators.

78. Internal fertilisation is the one in which syngamy _______________

79. Life begins in all sexually reproducing organisms from a <u>double-celled zygote</u>.
(*Rewrite the statement by changing the underlined word.*)

____

80. Write biological term for formation of embryo from zygote.

____

81. Choose the incorrect pair.
(i) Cell division in embryo – Increase the number of cells
(ii) Cell differentiation – Form specialised tissues and organs
(iii) Eggs covered by hard calcareous shell – Oviparous animals
(iv) Zygote develops outside the body –Viviparous animals

____

82. Chances of survival of young ones are more in the case of ________________ individuals.

83. Offspring of oviparous animals have less chances of survival as compared to those of viviparous animals because embryo does not develop completely.
(*Rewrite the statement by changing the underlined word.*)

____

84. Match the following columns.

	Column I		Column II
A.	Ovoviviparity	1.	Human beings
B.	Ovipary	2.	Bony fish
C.	Vivipary	3	Reptiles and birds

A. ____________ B. ____________ C. ____________

85. What does '*P*' signifies in the diagram given below?

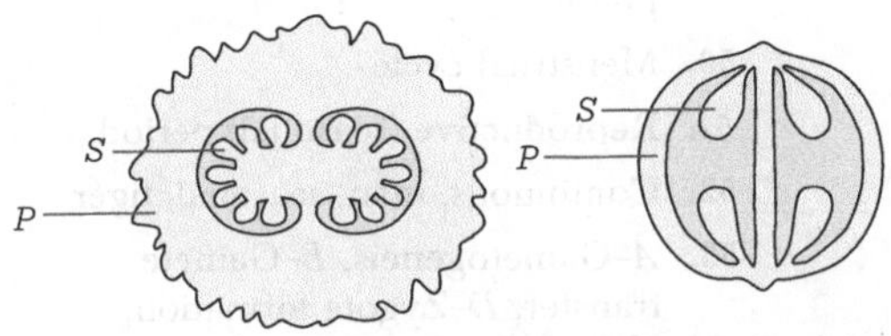

____

86. Match the following columns.

	Column I (Floral parts)		Column II (Converted to)
A.	Ovary	1.	Pericarp
B.	Ovule	2.	Perisperm
C.	Ovary wall	3.	Fruit
D.	Nucellus	4.	Seed

A. ____________ B. ____________ C. ____________ D. ____________

Answers

1. Lifespan
2. Shorter
3. *Amoeba* (single cell organism)
4. Reproduction
5. Rice plant → Banana tree → Rose bush → Banyan tree
6. (i) True
   (ii) False. Lifespan is not necessarily correlated with size and complexity of organisms.
7. Reproduction
8. Offspring
9. Birth → Growth → Death
10. True
11. A-4, B-1, C-2, D-3
12. *A*-70 years, *B*-Crow, *C*-Parrot *D*-12 years, *E*-20 years
13. 7 days, 5 weeks

14. False. Reproduction types in living organisms are based on participation of one organism or two organisms.
15. Genetically identical
16. Quick, one
17. Binary fission
18. Clone
19. *Amoeba* and *Paramecium*
20. Fusion
21. *A* – Bulbil of *Agave*
    B – Buds in *Hydra*
    C – Gemmules in Sponge
22. Budding
23. Oospores is odd one out because it is a sexual reproductive structure.
    Others are asexual reproductive structures.
24. Micropropagation
25. Fission
26. Encystation
27. Encystation, sporulation
28. (iii), gemmules are formed in sponges.
29. A-2, B-1, C-3
30. *A*-Conidia of *Penicillium*
    B-Adventitious buds of *Bryophyllum*
    C-Offset of water hyacinth
31. Plants
32. Sucker
33. Runner, rhizome, sucker, tuber, offset and bulb
34. Vegetative propagules
35. Fragmentation
36. True
37. Asexual
38. Nodes of the modified stem.
39. New plantlets in potato and sugarcane arise from bud on stem.
40. A-3, B-4, C-1, D-2
41. Algae and fungi
42. Sexual reproduction
43. Perennial
44. Juvenile phase
45. Juvenile phase → Pre-reproductive phase → Reproductive phase → Senescent phase.
46. (ii), bamboo lives vegetatively for 50-100 years.
47. 12
48. Juvenile phase (vegetative phase in plants)
49. False. The end of maturation phase marks the beginning of senescent phase.
50. Menstrual cycle
51. Reproductive (seasonal) period.
52. Continuous, man, seasonal, tiger
53. *A*–Gametogeneis, *B*–Gamete transfer, *D*–Zygote formation, *E*–Embryogenesis
54. Gametogenesis
55. *A*–Isogametes of *Cladophora*,
    B–Heterogametes of *Fucus*
    C–Heterogametes of human
56. Homogametes, heterogametes and anisogametes
57. Antherozoid and sperm
58. Bisexual or monoecious.
59. *A*-Oogonium (female sex organ)
    B-Antheridium (male sex organ)
60. Heterothallic, dioecious or unisexual
61. Staminate, Pistillate, *Cucurbita*, Papaya
62. True
63. Earthworm. Other three are unisexual.
64. Fish
65. Haploid
66. Gamete mother cell or meiocytes
67. Mitotic division.
68. (ii) is incorrect. Potato has 24 chromosome number in gamete and 48 chromosome number in meiocyte.
69. A-3, B-4, C-2, D-1
70. Fusion of gametes → Zygote → New individual (cell $2n$)
71. Self-fertilisation occurs in the bisexual flower.
72. All given organisms show self fertilisation except papaya because it is a dioecious plant which shows cross-fertilisation.
73. True
74. False. Fusion of male and female gametes is called syngamy or fertilisation.
75. Parthenogenesis
76. External and internal
77. Gametes, synchrony, syngamy, offspring.
78. Occurs inside the body
79. Single-celled zygote.
80. Embryogenesis
81. (iv) is incorrect because zygote develops inside the body in viviparous animals.
82. Viviparous
83. Proper embryonic care and protection is absent.
84. A-1, B-3, C-2
85. Thick pericarp
86. A-3, B-4, C-1, D-2

CHAPTER TWO

SEXUAL REPRODUCTION IN FLOWERING PLANTS

Topic 1 : Flower and Development of Gametes

1. All flowering plants show sexual reproduction.

   True ☐ False ☐

2. Identify the parts labelled as *A* to *G* in following figure.

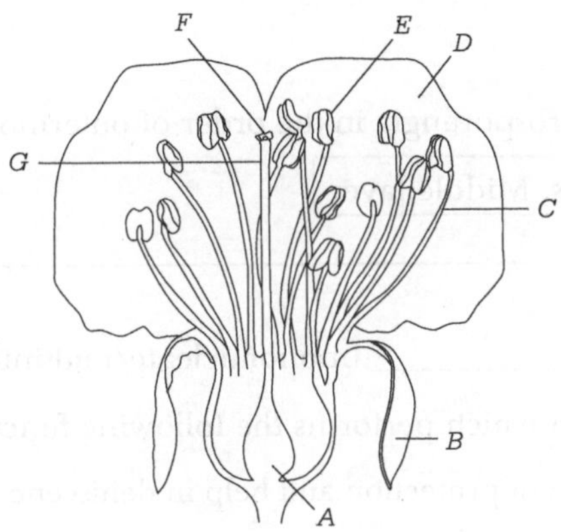

A. ____________________

B. ____________________

C. ____________________

D. ____________________

E. ____________________

F. ____________________

G. ____________________

3. Gynoecium represents the ____________________ (male/female) reproductive organ.

4. Arrange the following floral whorls on the thalamus from outermost to innermost.

Corolla, Androecium, Calyx, Gynoecium

   ____

5. Name the long, slender stalk of a typical stamen. ____________________

6. The <u>distal</u> end of the filament is attached to the thalamus or petal of the flower. (*Rewrite the statement by changing the underlined word.)*

____

7. Find odd one out.

Stamen, Stigma, Style, Ovary

____

8. The usually bilobed terminal structure of a typical angiospermic stamen is __________

9. Each lobe in a typical angiospermic anther is monothecous.

True ☐ False ☐

10. There are __________ (two/three/four) microsporangia in an anther.

11. After maturation, microsporangia develop into __________ .

12. Identify the parts labelled as *A* to *D* in the following diagram.

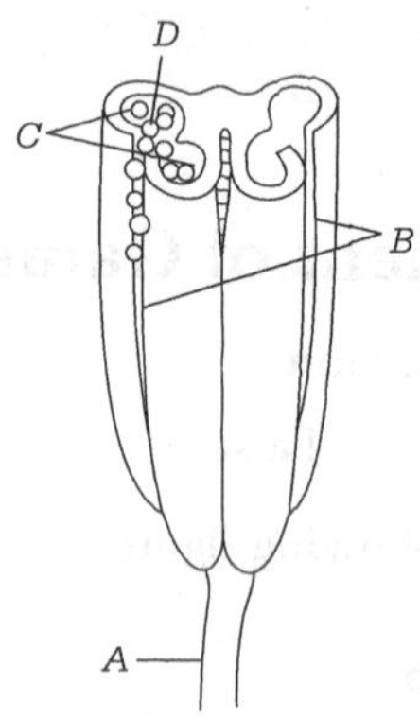

A. __________

B. __________

C. __________

D. __________

13. Arrange the following layers of microsporangia in the order of outermost to innermost.

Tapetum, Endothecium, Epidermis, Middle layer

____

14. Tapetum cells are __________ (mononucleate/multinucleate).

15. Name the layer of microsporangium which performs the following functions.

(i) Layers which perform the function of protection and help in dehiscence of anther to release the pollen.

____

(ii) Layer that nourishes the developing pollen grains.

____

16. Complete the following analogy.

Secretory tapetum : Glandular : : Plasmodial tapetum : __________

17. Select the correct words from the box given below to fill in the blanks.

Centre, Corner, Endogenous, Sporogenous, Homogeneous, Heterogeneous

A group of compactly arranged ______________ cells called the ______________ tissue occupies the ______________ of each microsporangium in a young anther.

18. The sporogenous tissue undergoes mitotic division to form microspore tetrads.

True ☐ False ☐

19. Identify the parts labelled as *A* to *E* in the following diagram.

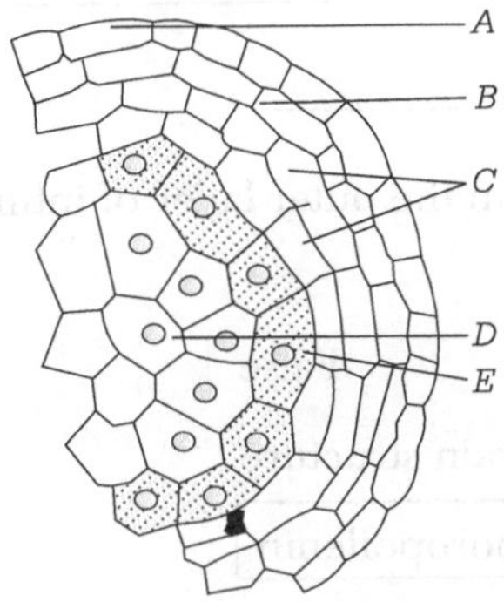

A. ______________

B. ______________

C. ______________

D. ______________

E. ______________

20. Identify the parts labelled as *A* to *F* in the following diagram.

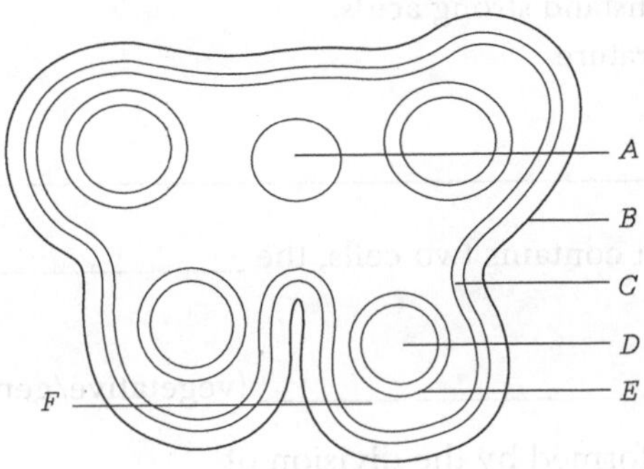

A. ______________

B. ______________

C. ______________

D. ______________

E. ______________

F. ______________

21. Write the expanded form of PMC. ______________

22. Name the process of formation of microspores from a PMC. ______________

23. Microspores are arranged in ______________ . As the anthers mature and dehydrate, microspores develop into the ______________ .

24. Match the following columns.

Column I (Parts of microsporangia)	Column II (Structural/Functional details)
A. Tapetum	1. Potential pollen
B. Anther	2. Released when anther dehisce
C. Pollen grains	3. Innermost layer
D. Sporogenous tissue	4. Bilobed

A.________ B.________ C.________ D. ________

25. Pollens have two prominent walls which are ________ and ________.

26. Exine is made up of ________.

27. Write the organic material which form the outer layer of intine. ________

28. Germ pores lack sporopollenin.

True ☐ False ☐

29. Find odd one out related to pollen grain structure.

Cellulose, Pectin, Hemicellulose, Sporopollenin

____

30. Read the following points and name the substance present in pollen grain.

(i) It is non-degradable and can withstand strong acids.
(ii) It is resistant at very high temperature.
(iii) No enzyme degrade it.

____

31. When the pollen grain is mature, it contains two cells, the ________ and ________.

32. Abundant food reserve is found in ________ (vegetative/generative) cells.

33. Male gametes in angiosperms are formed by the division of________.

34. Identify the structures marked *A* to *F* in the given diagram.

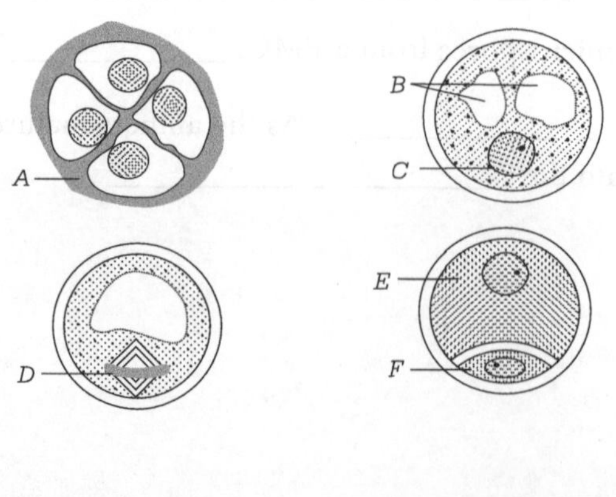

A. ________
B. ________
C. ________
D. ________
E. ________
F. ________

35. 60% of the angiosperms shed their pollens at the ____________________

36. Pollen grains cause allergies and chronic respiratory disorder in some people.

True ☐ False ☐

37. Write the scientific name of carrot grass. ____________________

38. ____________________ came into India as a contaminant causing pollen allergy.

39. Embryo sac consumption increases the performance of athletes.
(*Rewrite the statement by changing the underlined word.*)

____

40. Dendrogram is a schematic illustration of the pollen grain.
(*Rewrite the statement by changing the underlined word.*)

____

41. Honey is made by bees by digesting the pollen collected from flowers.

True ☐ False ☐

42. Find odd one out with respect to pollen viability of following members of family.

Rosaceae, Leguminosae, Solanaceae, Poaceae

____

43. Pollen grains can be stored for years in liquid nitrogen at 196°C.

True ☐ False ☐

44. Match the following columns.

Column I (Male gametophytes)	**Column II** (Structures)
A. Pollen	1. Irregular shaped nucleus
B. Exine	2. Race horses
C. Generative cell	3. Resistant to enzymes
D. Vegetative cell	4. Spindle-shaped, floating in cytoplasm

A. __________ B. __________ C. __________ D. __________

45. Fill in the blanks using words given in the box given below.

Multicarpellary, Monocarpellary, Androecium, Gynoecium

The ____________ representing female reproductive parts of flower may consist of single pistil, called ____________ or more than one pistil, called ____________ .

46. Identify whether the following diagrams are multicarpellary/monocarpellary and syncarpous/apocarpous.

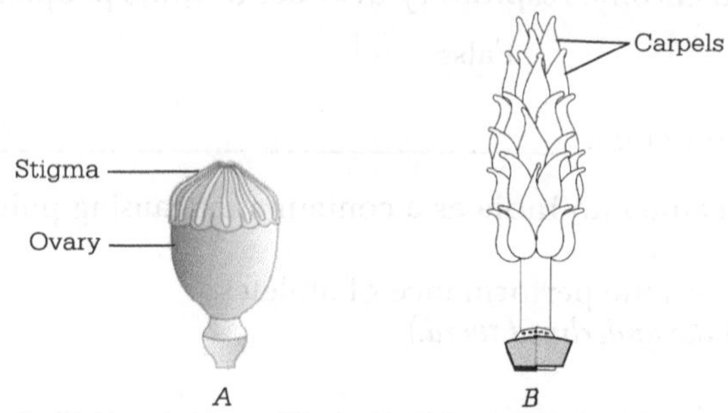

A. ______________________

B. ______________________

47. ______________________ is the landing platform for pollen grains.

48. Name the following.

(i) Basal bulged part of pistil ______________________

(ii) Elongated slender part beneath stigma ______________________

49. In the given diagram of pistil, in which part fertilisation takes place?

____

50. The placenta is located inside the ovarian cavity.

True ☐ False ☐

51. Write the other name of megasporangia. ______________________

52. Complete the following analogy with respect to number of ovules.

Wheat : ____________ : : Papaya : Many

53. Name the structure which connects ovule and placenta. ______________________

54. The junction between ovule and funicle is ______________________

55. The small opening at the tip of ovule where integument does not encircle is known as ______________________

56. Identify the parts labelled as *A* to *I* in the given diagram.

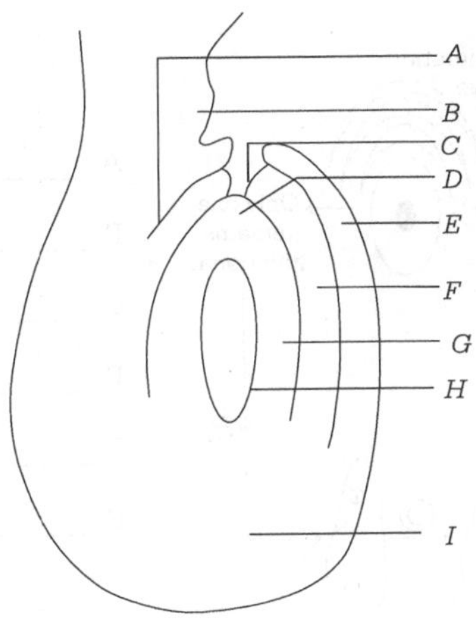

A. ____________________

B. ____________________

C. ____________________

D. ____________________

E. ____________________

F. ____________________

G. ____________________

H. ____________________

I. ____________________

57. Name the type of ovule which is inverted with body fused to funicle, micropyle lying close to hilum and facing placenta. ____________________

58. Each ovule has one or two protective envelopes called <u>chalaza</u>.
(*Rewrite the statement by changing the underlined word.*)

____

59. The mass of cells enclosed within the integument is called the nucellus.

True ☐ False ☐

60. Nucellus containing female gametophytes is also known as ____________________

61. Arrange the following structures in the order they appear from innerside to outerside of a typical ovule.

Nucellus, Outer integument, Inner integument, Embryo sac

____

62. Write the full form of MMC.____________________

63. The MMC undergoes ____________________ division, which results in the production ____________ megaspores. The formation of megaspore is ____________ .

64. If the endosperm mother cell of an angiosperm has 24 chromosomes, the same in megaspore mother cell would be ____________________

65. Fill in the blanks with words provided in the box given below.

Monosporic, Megaspore, Functional, Female, Embryo sac, Male, Disporic, Pollen tube

Only one __________ megaspore develops into __________ gametophyte and other three degenerate. This method of __________ formation from a single ______________ is termed as ____________ development.

66. Identify the parts labelled as A to F in the diagram given below.

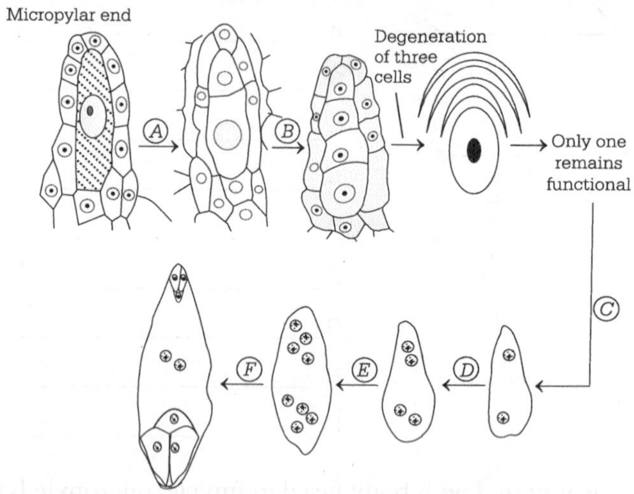

A. ____________________

B. ____________________

C. ____________________

D. ____________________

E. ____________________

F. ____________________

67. Identify the labelling parts of given diagrams.

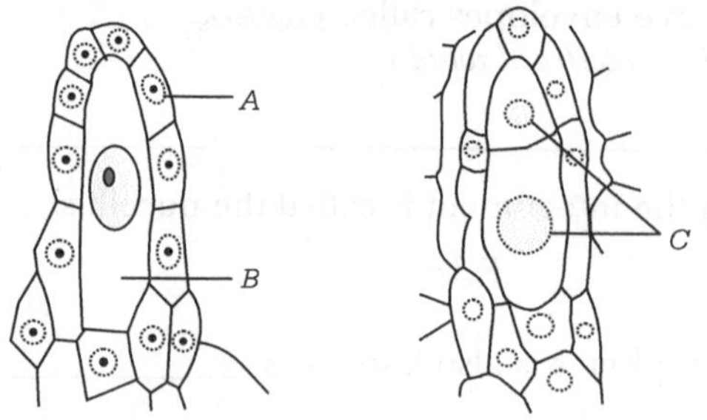

A. ____________________

B. ____________________

C. ____________________

68. Write the following terms in the sequence of embryo sac development.

Megaspore mother cell, Embryo sac, Megaspore, Archesporium

____

69. The archesporial cell divides by parietal division to form an outer ____________ and inner ____________ cell.

70. The two nuclei that is not surrounded by the cell wall in embryo sac is called ____________ nuclei.

71. Nuclear divisions are not followed immediately by cell wall formation.

True ☐ False ☐

72. The egg apparatus consists of two ____________ and one ____________.

73. Name the apparatus that is found at the micropylar tip, which guides the pollen tubes into the synergid. ____________________

74. Antipodals are the three cells at the ____________ end.

75. A typical angiosperm embryo sac is ________ (8/7) nucleate and ________ (6/7) celled.

76. Identify the parts labelled as *A* to *E* in the diagram given below.

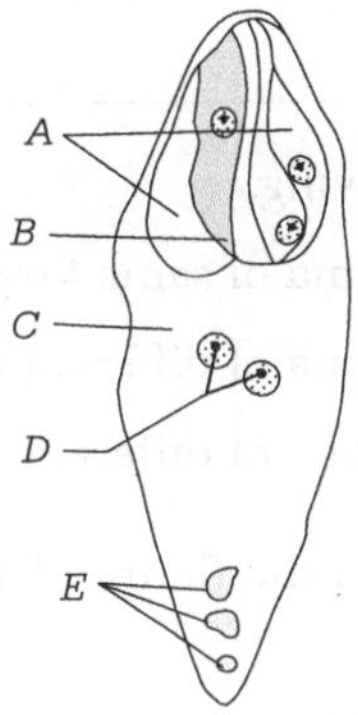

A. ____________

B. ____________

C. ____________

D. ____________

E. ____________

77. In embryo sac, the number of synergid, egg cell, central cell, antipodal cell follows the order.

____

78. Write the types of cell division (meiosis-I /meiosis-II/mitosis) involved in the following.

(i) Megaspore mother cell to megaspore dyad. ____________

(ii) Megaspore dyad to megaspore tetrad.____________

79. Complete the sequence.

Megaspore mother cell → (A) → Megaspore tetrad → Functional (B) → (C) Celled stage embryo sac

A.________ B.________ C.________

80. Antipodal cells commonly develop Ubisch bodies in embryo sac.

True ☐ False ☐

81. Identify the parts labelled as *A* to *F* in the diagram given below.

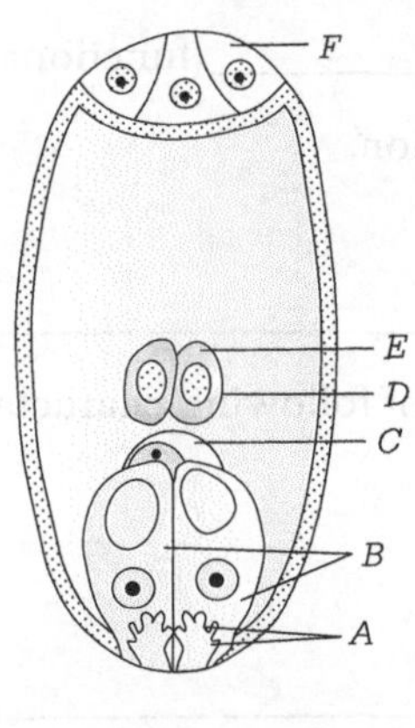

A. ____________

B. ____________

C. ____________

D. ____________

E. ____________

F. ____________

Topic 2 : Pollination

82. Mechanism of transfer of pollen grains to the ovule of a pistil is termed as pollination. (*Rewrite the statement by changing the underlined word.*)

____

83. Write the biological terms for the following.

(i) Transfer of pollen from anther to stigma of same flower. ____________

(ii) Transfer of pollen from anther to stigma of different plant. ____________

(iii) Transfer of pollen from anther to stigma of different flower on same plant. ____________

84. Identify the structures *A* and *B* shown in the figure of *Commelina* plant given below.

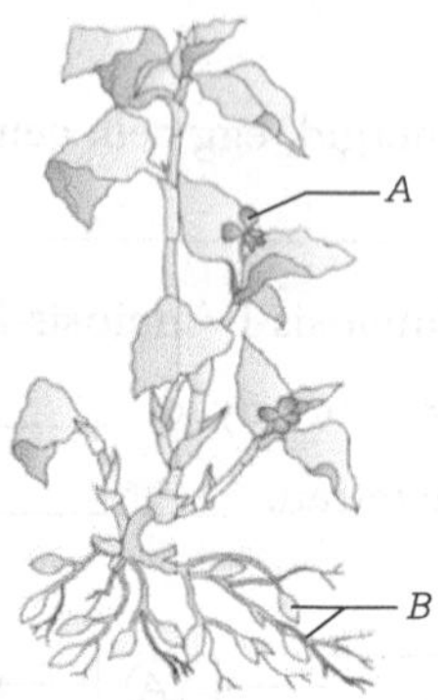

A. ____________

B. ____________

85. Cleistogamous flowers are strictly autogamous because they remain ____________

86. In chasmogamy, pollination takes place in ____________

87. Cleistogamous flowers produce assured seed set even in the absence of pollinators.

True ☐ False ☐

88. Geitonogamy is genetically similar to xenogamy.

True ☐ False ☐

89. Xenogamy causes ____________ (functionally/genetically) different flowers.

90. Find odd one out regarding pollination.

Wind, Light, Water, Insects

____

91. Which type of pollen grains consist of following characteristics?

(i) Non-sticky

(ii) Light

(iii) Produced in large number

____

92. Study the diagram given below about its structural details and write two characteristics of flowers which help in wind pollination.

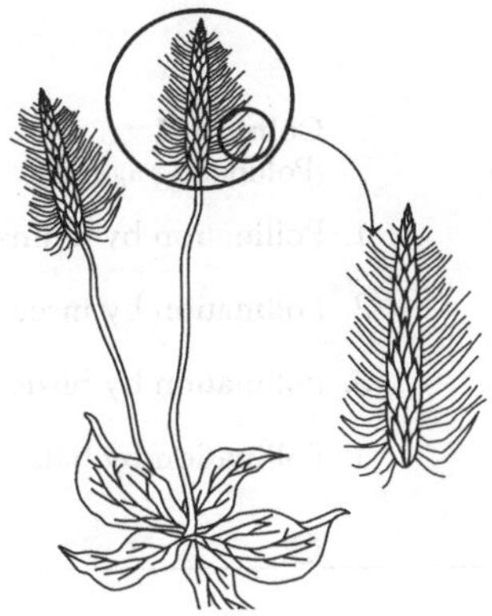

(i) ____________________ (ii) ____________________

93. Write any two examples of wind pollinated plants. ____________________

94. Complete the following analogy.

Hydrilla : ____________________ : : Water pollinated : Wind pollinated

95. In some water pollinated plants, pollens are protected from wetting by a ____________ covering.

96. Long, ribbon-like pollens are commonly seen in water pollinated plants.

True ☐ False ☐

97. Match the following columns.

Column I (Pollinating agents)	**Column II** (Names of floral species)
A. Lemurs	1. *Adansonia*
B. Gecko lizard	2. Traveler's palm
C. Bats	3. Mulungu

A.____________ B.____________ C.____________

98. ____________________ are the dominant biotic pollinating agents.

99. Identify the pollinators of the following flowers.

(i) Colourful, fragnant flowers ____________________

(ii) Foul odour flowers ____________________

100. Write the two floral rewards for the animal pollinators. ____________________

101. Write the name of plants which provide following floral rewards.

(i) Safe place to lay eggs. ____________________

(ii) Deposition of eggs in the locule of ovary by a moth.____________________

102. Write the common terms used for insects consuming pollen or nectar without bringing about pollination. ______

103. Match the following columns.

Column I (Types of pollination)	Column II (Pollinating agents)
A. Ornithophily	1. Pollination by animals
B. Zoophily	2. Pollination by insects
C. Chiropterophily	3. Pollination by birds
D. Entomophily	4. Pollination by bats

A.______ B. ______ C.______ D. ______

104. Which one of the following plant shows a very close relationship with a species of moth, where none of the two can complete its life cycle without the other?

Banana, *Yucca, Hydrilla, Viola*

______

105. Continued self-pollination results in______

106. Write any two ways by which flowering plants prevent self-pollination.

______

107. Name the genetic mechanism of inhibiting pollen germination or pollen tube growth in the pistil. ______

108. Identify the incorrect pairs.

(i) Castor - Monoecious

(ii) Maize - Dioecious

(iii) Papaya - Dioecious

______

109. A dioecious flowering plant prevents both ______

110. In which of the following, both autogamy and geitonogamy are prevented?

Wheat, Papaya, Castor, Maize

______

111. Monoecious plants prevent geitonogamy, but not autogamy. (*Rewrite the statement by changing the underlined words.*)

______

112. Acceptance or rejection of a pollen is a result of continuous dialogue between pollen grain and the pistil.

True ☐ False ☐

113. Arrange the following events of pollen grain germination in a sequence.

(i) Pollen tube growth through stigma tissues.

(ii) Shedding of pollen grains at 2/3 celled condition.

(iii) Entry of pollen into the ovule through micropyle.

(iv) Pollen grain germination on the stigma.

____

114. Identify the parts labelled as *A* to *F* in the given diagram.

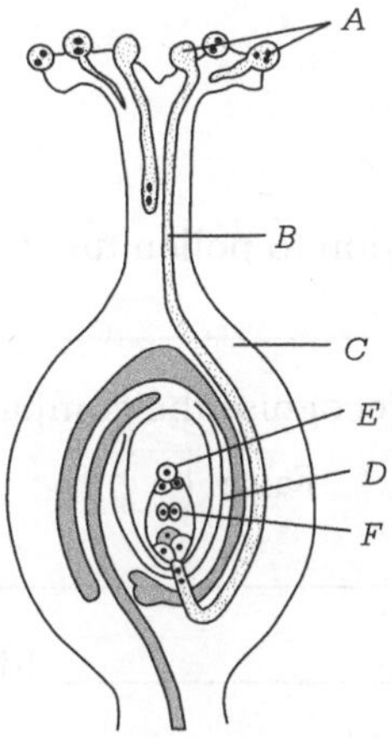

A. ____________________

B. ____________________

C. ____________________

D. ____________________

E. ____________________

F. ____________________

115. **Diagram showing entry of pollen tube to the embryo sac. Identify the parts labelled as *A* to *G* in the diagram given below.**

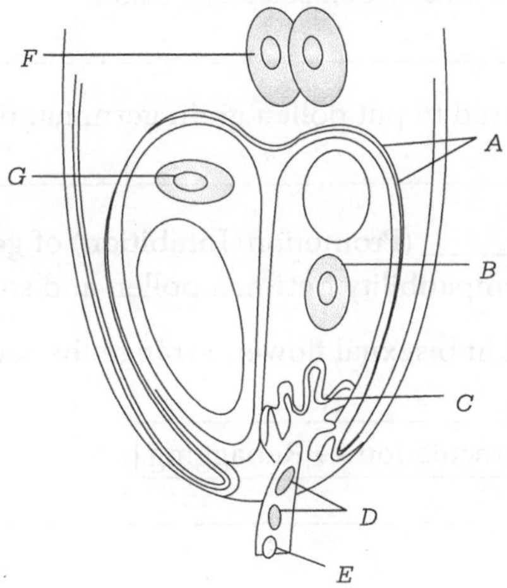

A. ____________________

B. ____________________

C. ____________________

D. ____________________

E. ____________________

F. ____________________

G. ____________________

116. **The ____________________ cell divides and forms the two male gametes during pollen tube growth in 2-celled pollen grains.**

117. Diagram showing discharge of gametes in the egg apparatus. Identify the parts labelled as *A* and *B*.

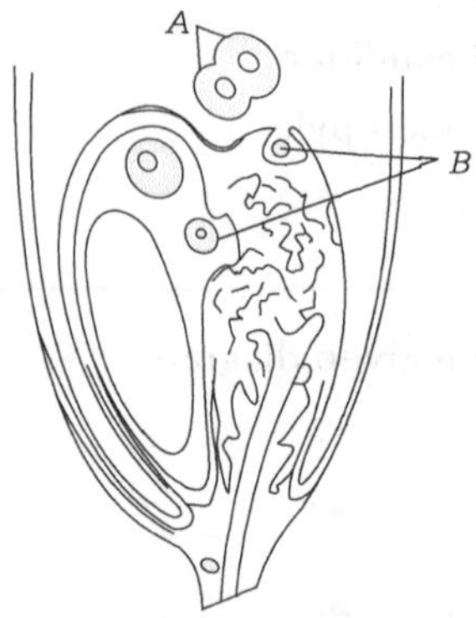

A. ______________________

B. ______________________

118. Name the interaction from pollen deposition to pollen tubes entry to the ovule.

____

119. The pistil alone does not have ability to recognise the compatibility of pollen.

True ☐ False ☐

120. Pollen pistil recognition promotes the ______________ events.

121. Pollen-pistil interaction is a ______________ (dyanamic/not worthy) process in the hybridisation technique.

122. ______________ is the effective pollination between flowers of different species of the same genus or between different genera.

123. Removal of anthers from flower before anther dehiscence is called

____

124. Give reason why 10% sugar solution used to put pollen grain germination on glass slide.

____

125. ______________ (Promotion/Inhibition) of germ tube from pollen occurs when there is genetic compatibility between pollen and stigma.

126. For artificial hybridisation experiment in bisexual flower, arrange the sequence in correct order?

Cross-pollination → Bagging → Emasculation → Rebagging

____

Topic 3 : Double Fertilisation and Development of Endosperm

127. Pollen tube releases ____________ (one/two/three) male gametes into the cytoplasm of synergid.

128. Zygote is a triploid structure .

True ☐ False ☐

129. Name the structure formed by the fusion of a male gamete and polar nuclei in the central cell.____________

130. Write the ploidy of the PEN ____________

131. Write the number of chromosomes of the following if the haploid number in a flowering plant is 14.

(i) Integuments____________

(ii) Embryo____________

(iii) Endosperm____________

132. PEN formation involves fusion of three ____________ (haploid/diploid) nuclei and therefore it is called ____________ fusion.

133. The fertilisation in flowering plants is termed as ____________ fertilisation as it involves two types of fusion, namely ____________ and ____________ .

134. The fusion of one male gamete with egg cell is called ____________ (xenogamy/syngamy).

135. Identify the parts labelled as *A*, *B* and *C*.

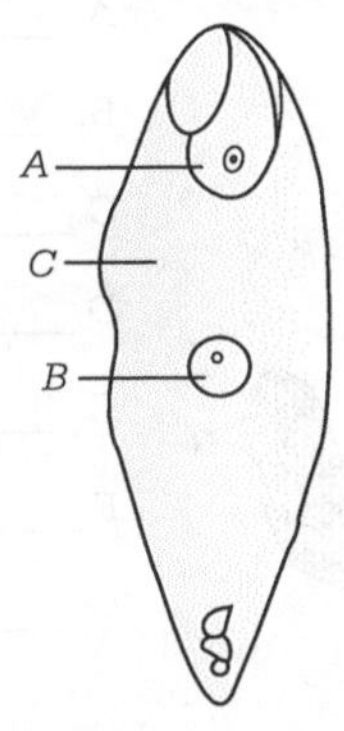

A. ____________

B. ____________

C. ____________

136. If endosperm has 36 number of chromosomes then find out the chromosome number of male and female gamete.____________

137. In an angiosperm, male plant is diploid and female plant is tetraploid then endosperm will be.____________

138. In angiosperm, pollen tube librates their male gametes into the ____________

139. Coconut water from a tender coconut is ____________

140. Endosperm is consumed by developing embryo in the seed of ________ (pea/castor).

141. Plumule is produced from the apical octant.

True ☐ False ☐

142. Post-fertilisation events do not include seed and fruit formation.

True ☐ False ☐

143. ____________ (Embryo/Endosperm) development precedes ____________ (endosperm/embryo) development.

144. The developing embryo derives nutrition from ____________ tissue.

145. Aleurone cells are present in endosperm of cereals.

True ☐ False ☐

146. Write the stage of endosperm development where PEN undergoes successive nuclear division to give rise to free nuclei. ____________

147. The white coconut kernel is made up of ____________

148. Write any two seeds in which endosperm is completely consumed by the developing embryo. ____________

Topic 4 : Embryo, Seed and Fruit Formation

149. Identify the different stages of embryogenesis (*A* to *G*) in the given diagram.

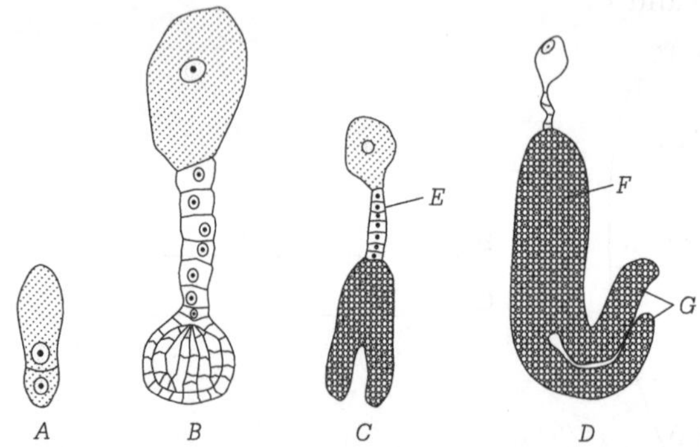

A. ____________

B. ____________

C. ____________

D. ____________

E. ____________

F. ____________

G. ____________

150. The study of embryo development is known as ____________

151. The early stage of embryo development is called somatogeny.

True ☐ False ☐

152. Embryogeny are different in both monocots and dicots.

True ☐ False ☐

153. Write the following words in the sequence of development.

Heart-shaped embryo, Mature embryo, Globular embryo, Endosperm development, Proembryo.

____

154. Identify the parts labelled *A* to *E* in the following diagram of typical dicot embryo.

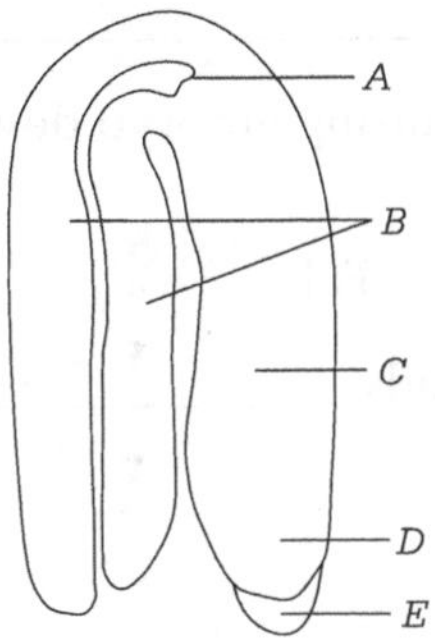

A. ____________

B. ____________

C. ____________

D. ____________

E. ____________

155. In the figure given below, find out the type of seed and identify which labelled parts represent coleoptile, coleorhiza and epiblast.

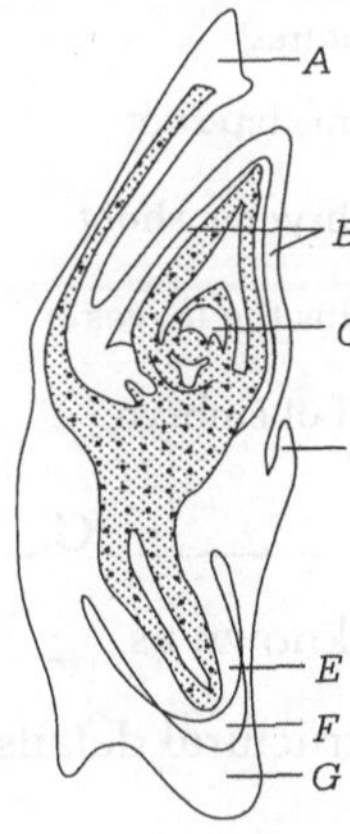

____

156. A typical dicotyledonous embryo consists of an ____________ axis and ____________ cotyledons. The portion of embryonal axis above the level of cotyledons is ____________ which terminates with the ____________ or stem tip.

157. Complete the following analogy.

Epicotyl : ____________ : : Plumule : Radicle

158. Name the structure covering the radicle. ____________

159. Write other name for cotyledons used in grass family. ____________

160. The cylindrical portion below the cotyledons is ____________ .

161. The sheath covering the radicle and root cap in monocots is known as ____________

162. Name the foliar hollow structure enclosing leaf primordia and a shoot apex. ________

163. Seed is a fertilised ovary.

True ☐ False ☐

164. Find the odd one out.

Seed coat, Cotyledons, Pericarp, Embryo, Axis

____

165. Endosperm is not completely consumed during embryo development in non-albuminous seeds.

True ☐ False ☐

166. Write two examples of the albuminous seeds. ________________

167. Find the odd one out.

Castor, Bean, Wheat, Rice

____

168. Match the following columns.

	Column I (Parts of seed)		Column II (Features)
A.	Cotyledon	1.	Forms tap root
B.	Epicotyl	2.	Embryonic shoot
C.	Plumule	3.	Forms the leaves
D.	Hypocotyl	4.	Leaf of embryo

A.__________ B.__________ C.__________ D. __________

169. The persistent residual nucellus is known as ________________

170. The diagram given below shows structural details of a dicot seed. Study the diagram and answer the quesitons that follows.

(i) Identify the structures labelled as *A* to *E*.

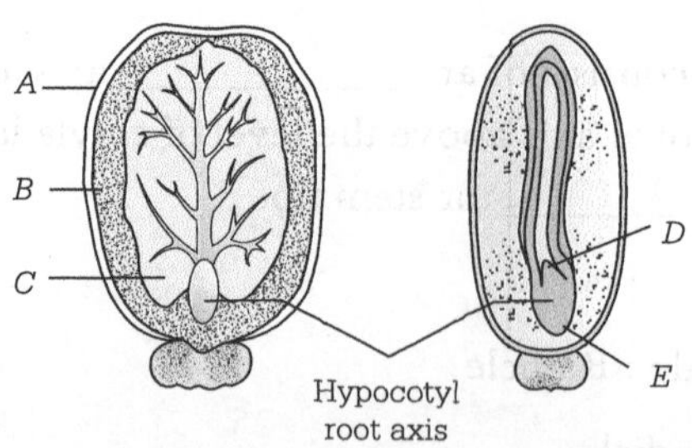

A. ______________

B. ______________

C. ______________

D. ______________

E. ______________

(ii) Name the plant from which this seed is obtained. ________________

(iii) Write the function of seed parts labelled as *A* and *B*.

____

171. Identify the parts labelled as *A* to *D* in structure of seed given below.

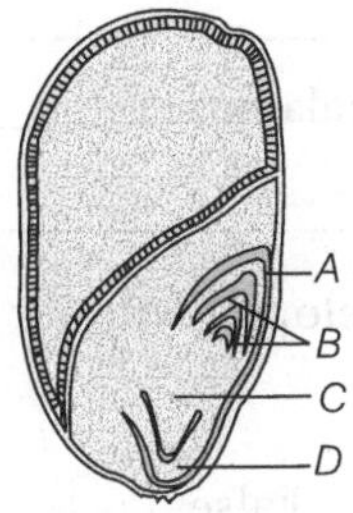

A. ______________________

B. ______________________

C. ______________________

D. ______________________

172. Match the following columns.

	Column I (Seeds / Fruits)		Column II (Examples)
A.	Parthenocarpic fruit	1.	Banana
B.	True fruit	2.	Mango
C.	False fruit	3.	Cashew

A.______________ B.______________ C.______________

173. The seed shows dormancy when seed is ______________ and its seed coat is ________ .

174. *Phoenix dactylifera* (date palm) is an example of showing ______________ years old seed to be viable.

175. What is common seed characteristics of the plants given in the box.

Orobanche, Striga, *Ficus*, Orchids

____

176. The wall of <u>ovule</u> develops into wall of the fruit.
(*Rewrite the statement by changing the underlined word.*)

____

177. The wall of fruit is known as ______________________

178. Match the following columns.

	Column I		Column II
A	Ovule	1.	Fruit
B.	Ovary	2.	Orange
C.	Dry fruit	3.	Mustard
D.	Fleshy fruit	4.	Seed

A.______________ B.______________ C .______________ D.______________

179. Give a common term for the following.
(i) Fruits developing from ovary only. ____________________
(ii) Fruits developing from both ovary and thalamus. ____________________

180. Write two examples of false fruits. ____________________

181. Write the scientific name for the fruits developed without fertilisation. ____________________

182. Parthenocarpic fruits are seedless.
True ☐ False ☐

183. Parthenocarpy can be induced by ____________________.

184. Seeds generate new genetic combinations.
True ☐ False ☐

185. ____________ and ____________ of mature seeds are important for storage of seeds and hence in the agricultural field.

186. Write the scientific name of the plant having oldest viable seed excavated from Arctic Tundra. ____________________

Topic 5 : Apomixis and Polyembryony

187. Name the mechanism by which seeds are produced without fertilisation. ____________

188. Asteraceae and grasses usually adapt apomixis.
True ☐ False ☐

189. ____________ is an asexual reproduction that mimics sexual reproduction.

190. Give the name for the occurrence of more than one embryo in a seed.
____

191. Write two examples of fruits exhibiting polyembryony. ____________________

192. Find odd one out with respect to fruit development.

Orange, Papaya, Apple, Pomegranate

____

193. Formation of diploid egg cell without reduction division is the way of ____________ (apomictic/parthenocarpic) seed development in same species.

194. Cultivation of hybrids has tremendously ____________ (increased/decreased) productivity.

195. Seeds collected from hybrids do not maintain ____________ characters.

196. There is segregation of characters in the hybrid progeny when the hybrids are made into apomicts.
True ☐ False ☐

197. Name the cell which protrude into the embryo sac and develop into the embryo during polyembryony.
____

Answers

1. True
2. *A*–Ovary, *B*–Sepal, *C*–Filament, *D*–Petal, *E*–Anther, *F*–Stigma, *G*–Style
3. Female
4. Calyx → Corolla → Androecium → Gynoecium
5. Filament
6. Proximal
7. Stamen is odd one out as it represents male reproductive part of flower and rest others are female reproductive parts of flower.
8. Anther
9. False. Each lobe in a typical angiospermic anther is dithecous.
10. Four
11. Pollen sacs
12. *A*–Filament (stalk), *B*–Line of dehiscence, *C*–Pollen sacs, *D*–Pollen grain
13. Epidermis → Endothecium → Middle layer → Tapetum.
14. Multinucleate
15. (i) Epidermis, endothecium and middle layer.
    (ii) Tapetum
16. Amoeboid
17. Homogeneous, sporogenous, centre
18. False. The sporogenous tissue undergoes meiotic division to form microspore tetrads.
19. *A*–Epidermis, *B*–Endothecium, *C*–Middle layers, *D*–Microspore mother cell, *E*–Tapetum
20. *A*–Connective, *B*–Epidermis, *C*–Endothecium, *D*–Sporogenous tissue, *E*–Tapetum, *F*–Middle layer
21. Pollen Mother Cell
22. Microsporogenesis
23. Microspore tetrad, pollen grains
24. A-3, B-4, C-2, D-1
25. Exine, intine
26. Sporopollenin
27. Cellulose and pectin
28. True
29. Hemicellulose. Rest of the options are the constituents of exine and intine.
30. Sporopollenin
31. Vegetative cell, generative cell
32. Vegetative
33. Generative cell
34. *A*–Pollen tetrad, *B*–Vacuoles, *C*–Nucleus, *D*–Asymmetric spindle, *E*–Vegetative cell, *F*–Generative cell
35. 2-celled stage
36. True
37. *Parthenium*
38. Carrot grass
39. Pollen
40. Palenogram
41. True
42. Poaceae is odd one out. Rest of the options are the plant families which can maintain pollen viability for months whereas Poaceae member like rice lose their pollen viability within 30 minutes of their release.
43. False. Pollen grains can be stored for years in liquid nitrogen at −196°C.
44. A-2, B-3, C-4, D-1
45. Gynoecium, monocarpellary, multicarpellary
46. *A*-Multicarpellary and syncarpous
    B-Multicarpellary and apocarpous.
47. Stigma
48. (i) Ovary, (ii) style
49. *C*-Ovary
50. True
51. Ovule
52. One
53. Funicle
54. Hilum
55. Micropyle
56. *A*–Hilum, *B*–Funicle, *C*–Micropyle, *D*–Micropylar pole, *E*–Outer integuments, *F*–Inner integument, *G*–Nucellus, *H*–Embryo sac, *I*–Chalazal pole
57. Anatropous ovule
58. Integuments
59. True
60. Embryo sac
61. Embryo sac → Nucellus → Inner integument→ Outer integument
62. Megaspore Mother Cell
63. Meiotic, four, megasporogenesis
64. 16
65. Functional, female, embryo sac, megaspore, monosporic.
66. *A*–Meiosis-I, *B*– Meiosis-II, *C*–Mitosis, *D*–Mitosis, *E*–Mitosis, *F*– Cytokinesis
67. *A*–Nucellus, B–Megaspore Mother Cell (MMC), *C*–Megaspore dyad
68. Archesporium → Megaspore → Megaspore mother cell → Embryo sac
69. Parietal, sporogenous
70. Polar
71. True
72. Synergids, egg cell.
73. Filiform apparatus
74. Chalazal
75. 8, 7.
76. *A*–Synergids, *B*–Egg, *C*–Central cell, *D*–2 polar nuclei, *E*–Antipodal cells
77. 2-1-1-3
78. (i) Meiosis-I
    (ii) Meiosis-II
79. *A*-Megaspore dyad, *B*-Megaspore,
    C-8
80. False. The antipodal cells commonly develop houstoria in embryo sac.

81. *A*–Filiform apparatus, *B*–Synergids, *C*–Egg, *D*–Central cell, *E*–Polar nuclei, *F*–Antipodal cells
82. Stigma
83. (i) Autogamy, (ii) Xenogamy, (iii) Geitonogamy.
84. *A*-Chasmogamous flower
B-Cleistogamous flower
85. Always close
86. Open flower
87. True
88. False. Geitonogamy is genetically similar to autogamy.
89. Genetically
90. Light is odd one out. Rest of the options are agents of pollination.
91. Wind pollinated pollen grains.
92. (i) Large feathery stigma
(ii) Compact inflorescence.
93. Corn, wheat
94. Corn or grass
95. Mucilaginous
96. True
97. A-2, B-3, C-1
98. Bees
99. (i) Bees, (ii) Flies/Beetles
100. Nectar and pollen grains
101. (i) *Amorphophallus*, (ii) *Yucca*
102. Pollen/nectar robbers
103. A-3, B-1, C-4, D-2
104. *Yucca*
105. Inbreeding depression
106. (i) Non-synchronisation between pollen release and stigma receptivity.
(ii) Placement of anther and stigma at different positions.
107. Self-incompatibility
108. (ii) is incorrect because maize is monoecious
109. Cleistogamy and xenogamy
110. Papaya
111. Autogamy, geitonogamy.
112. True
113. (ii) → (iv) → (i) → (iii)
114. *A*–Pollen grains, *B*–Pollen tube, *C*–Ovary, *D*–Ovule, *E*–Antipodal cell, *F*– Polar nuclei
115. *A*–Plasma membrane, *B*–Synergid, *C*–Filiform apparatus, *D*–Male gametes, *E*–Vegetative nucleus, *F*–Central cell, *G*–Egg nucleus
116. Generative
117. *A*–Polar nuclei, *B*–Male gametes
118. Pollen-pistil interaction
119. False. The pistil alone have the ability to recognise the compatible and incompatible pollen.
120. Post-pollination.
121. Dyanamic
122. Artificial hybridisation
123. Emasculation
124. Sugar solution provides nutrition to germinating pollen.
125. Promotion
126. Emasculation → Bagging → Cross-pollination → Rebagging.
127. Two
128. False. Zygote is a diploid structure.
129. Primary Endosperm Nucleus (PEN)
130. Triploid
131. (i) 28, i.e. $2n$
(ii) 28, i.e. $2n$
(iii) 42, i.e. $3n$
132. Haploid, triple
133. Double, syngamy, triple fusion
134. Syngamy
135. *A*-Zygote, *B*- Primary Endosperm Nucleus (PEN), *C*-Primary Endosperm Cell (PEC)
136. Each of them have 12 number of chromosome.
137. Pentaploid
138. Synergids
139. Free-nuclear endosperm
140. Pea
141. True
142. False. Post-fertilisation events include seed and fruit formation.
143. Endosperm, embryo
144. Endosperm
145. True
146. Free-nuclear endosperm stage
147. Cellular endosperm
148. Orchid, groundnut
149. *A*–2-celled stage, *B*–Globular embryo, *C*–Heart-shaped embryo, *D*–Mature embryo, *E*–Suspensor, *F*–Radicle, *G*–Cotyledon
150. Embryology
151. False. The early stage of embryo development is known as embryogeny.
152. False. The embryogeny are similar in both monocots and dicots.
153. Endosperm development → Proembryo → Globular embryo → Heart-shaped embryo → Mature embryo
154. *A*–Plumule, *B*–Cotyledons, *C*–Hypocotyl, *D*–Radicle, *E*–Root cap
155. Monocot, *B*–Coleoptile, *D*–Epiblast, *G*–Coleorhiza
156. Embryonal, two, epicotyl, plumule
157. Hypocotyl
158. Root cap
159. Scutellum
160. Hypocotyl
161. Coleorhiza
162. Coleoptile
163. False. Seed is a fertilised ovule.
164. Pericarp is odd one out. Rest of the options are parts of a seed. It is a part of fruit.
165. False. Endosperm is not completely consumed during embryo development in albuminous seeds.
166. Barley, castor
167. Bean is odd one out. Rest of the plants have endospermic seeds.

168. A-4, B-2, C-3, D-1

169. Perisperm

170. (i) *A*-Seed coat
B-Endosperm
C-Cotyledon
D-Shoot apical meristem
E-Root tip
(ii) Castor (*Ricinus communis*)
(iii) *A*-Seed coat, protects the internal structures of seed.
B-Endosperm provides nutrition to the developing embryo.

171. *A*–Coleoptile, *B*–Plumules, *C*–Radicle, *D*–Coleorhiza

172. A-1, B-3, C-2

173. Dehydrated, hard.

174. 2000

175. They all produce huge number of seeds.

176. Ovary

177. Pericarp

178. A-4, B-1, C-3, D-2

179. (i) True fruits
(ii) False fruits

180. Apple and strawberry

181. Parthenocarpic fruit

182. True

183. Growth hormone

184. True

185. Dehydration, dormancy

186. *Lupinus arcticus*

187. Apomixis

188. True

189. Apomixis

190. Polyembryony

191. Orange, mango

192. Apple is odd one out. Rest of the options are polyembryonic fruits.

193. Apomictic

194. Increased

195. Hybrid

196. False. There is no segregation of characters in the hybrid progeny when the hybrids are made into apomicts.

197. Nucellar cells.

CHAPTER THREE

HUMAN REPRODUCTION

Topic 1 : Male Reproductive System

1. Name the pouch in which testes are suspended outside the abdominal cavity. ____________________

2. The ____________ (scrotum/hypothalamus) helps in the maintenance of low temperature (~ 34-35°C) necessary for spermatogenesis.

3. Answer the following questions in numbers only.
   (i) The length of each testis. ____________
   (ii) The width of each testis. ____________
   (iii) The number of testicular lobules in each testis. ____________
   (iv) The number of seminiferous tubules in each lobule. ____________

4. The seminiferous tubules of the testis are lined on its inside by spermatids and Sertoli cells. (*Rewrite the statement by changing the underlined word.*)

   ____

5. ____________ which provide nutrition to the developing of male germ cells.

6. Which of the following cells present in the mammalian testes form the sperms?

Leydig cells, Spermatogonia, Interstitial cells, Sertoli cells

   ____

7. The region outside the seminiferous tubules are called interstitial spaces contain small blood vessels and masses of cells called ____________

8. The seminiferous tubules of the testis open into the vasa efferentia by rete testis.

   True ☐ False ☐

9. Complete the sequence for transport of sperm cells in male reproductive system.
Seminiferous tubules → __________ → Vasa efferentia → Epididymis → __________ → Ejaculatory duct → Urethra → __________ .

10. Choose the incorrect pair.
(i) Leydig cells – Secrete LH and FSH
(ii) Vasa efferentia and epididymis – Accessory ducts
(iii) Vas deferens – Loops over the urinary bladder

____

11. Choose the correct option.

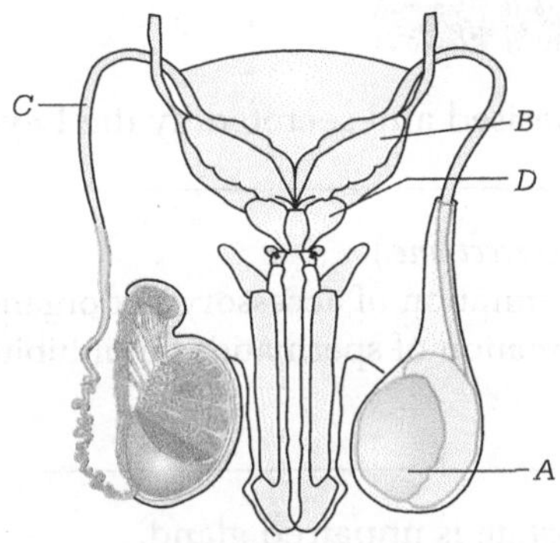

(i) *A*– Testis–possesses 3-4 testicular lobules
(ii) *B*– Seminal vesicle–storage of sperm
(iii) *C*– Vas deferens–helps in sperm transfer
(iv) *D*– Prostate gland–secretes seminal fluid

____

12. The shared terminal duct of the reproductive and urinary system in the human male is

____

13. Identify the labelled parts as *A, B, C* and *D* in the given diagram.

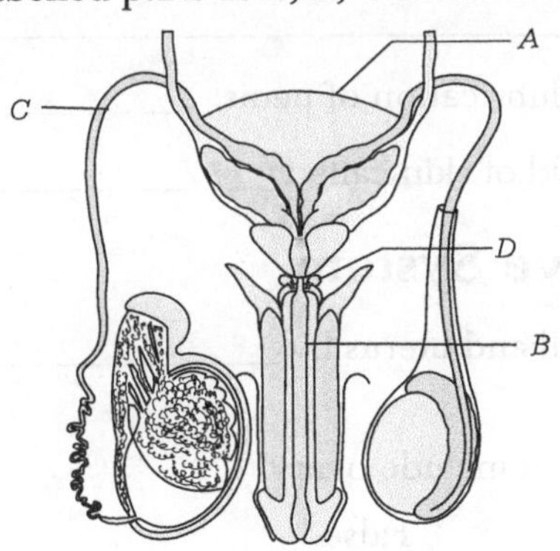

A. __________
B. __________
C. __________
D. __________

14. Identify the accessory reproductive gland found in males.

Seminal vesicles, Testis, Hypothalamus

____

15. The given diagram refers to TS of testis showing sectional view of a few seminiferous tubules. Identify the labelled parts as *A-D*.

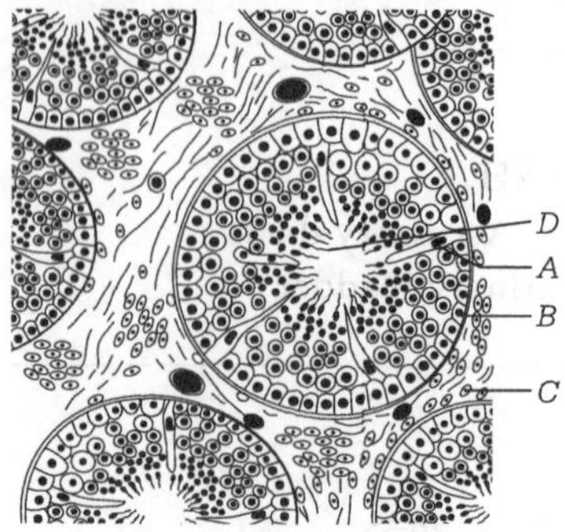

A. ____________

B. ____________

C. ____________

D. ____________

16. Name the testicular hormone synthesised and secreted by the Leydig cells.

____

17. Testosterone in males (*Choose the incorrect one.*)
   (i) regulates the development and maturation of accessory sex organs.
   (ii) plays a stimulatory role in the formation of spermatids but inhibitory role in spermiogenesis.

____

18. Seminal vesicle is paired, while prostate is unpaired gland.

True ☐ False ☐

19. Find odd one out.

Rete testis, Vasa efferentia, Epididymis, Vas deferens, Seminiferous tubule

____

20. Write the correct options from the following. The constituents of seminal plasma are

Fructose, Glucose, Calcium, Enzymes

____

21. Name the gland responsible for the lubrication of penis. ____________

22. Glans penis is covered by a loose fold of skin called ____________

Topic 2 : Female Reproductive System

23. Ovary is connected to the pelvic wall and uterus by ____________ (ligaments/ovarian stroma).

24. The female external genitalia does not include ovary.

True ☐ False ☐

25. The ovaries of human females (*Choose the correct one.*)
   (i) are the primary female sex organs.
   (ii) produce FSH and LH.

____

26. Give a collective term for oviducts, uterus and vagina. ______________________

27. The following diagram refers to female reproductive system of human. Identify the labelled parts as *A* to *E*.

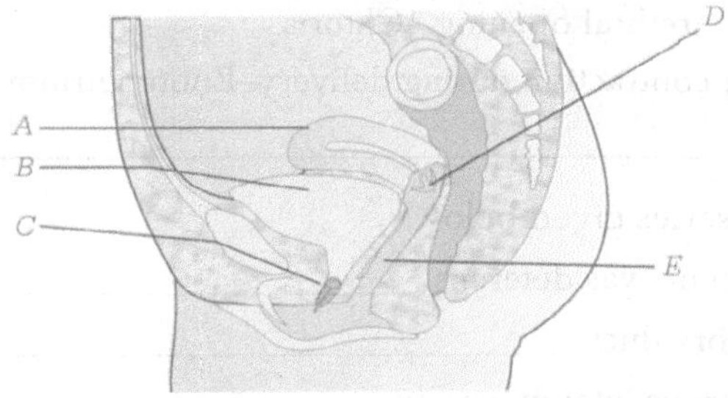

A. ______________________

B. ______________________

C. ______________________

D. ______________________

E. ______________________

28. Consider the figure given below and identify the parts which are incorrectly labelled.

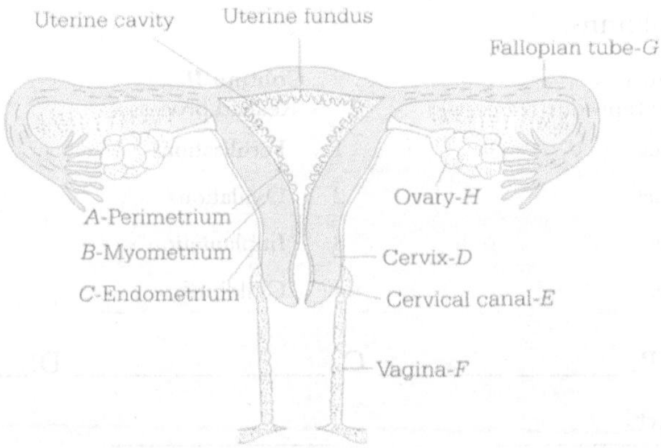

____

29. Write the scientific name for the following.

(i) Finger-like projections on the edges of infundibulum. ______________________

(ii) Last part of oviduct with a narrow lumen joining the uterus. ______________________

(iii) The part through which uterus opens into the vagina. ______________________

30. Name the wall of the uterus which sloughs off during menstruation.

____

31. Find the odd one out.

Mons pubis, Pectoralis major, Labia majora, Labia minora

____

32. The membrane which partially covers the opening of vagina is known as <u>clitoris</u>.
(*Rewrite the statement by changing the underlined word.*)

____

33. The presence or absence of hymen is a reliable indicator of virginity or sexual experience.

True ☐ False ☐

34. Choose the incorrect pair.

(i) Cushion of fatty tissue covered by pubic hair –Mons pubis

(ii) Single, inverted pear–Womb-shape structure

(iii) Finger-like structure above the urethral opening –Clitoris

(iv) Uterine layer exhibiting strong contraction during delivery–Endometrium

____

35. Pick the odd one out from each series given below.

I. Scrotum, rete testis, Fallopian tube, vas deferens. ______________

II. Ovary, uterus, vagina, ejaculatory duct. ______________

III. Acrosome, Graafian follicle, corpus luteum, cervix. ______________

IV. Prostate, testis, seminal vesicles, Cowper's gland. ______________

36. Match the following columns.

	Column I (Female reproductive parts)		Column II (Related processes)
A.	Ovaries	1.	Fertilisation
B.	Oviduct	2.	Ovulation
C.	Uterus	3.	Implantation
D.	Cervix	4.	Childbirth

A. ______________ B. ______________ C. ______________ D. ______________

37. Complete the flow chart.

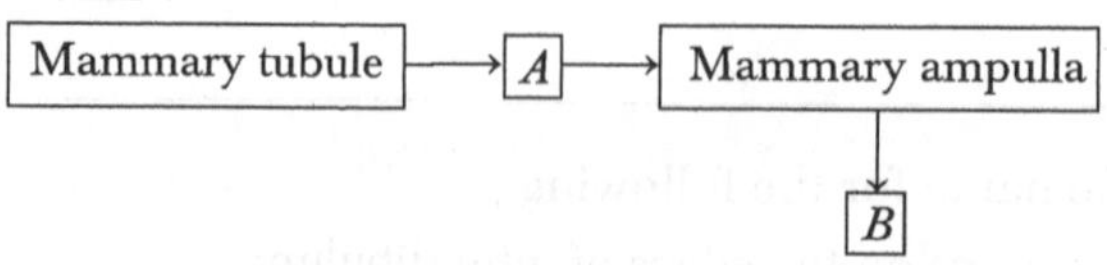

A. ______________ B. ______________

38. The main tissue present in breast is ______________ tissue.

39. Given the diagrammatic sectional view of mammary gland. Identify the labelled parts as *A*, *B*, *C* and *D*.

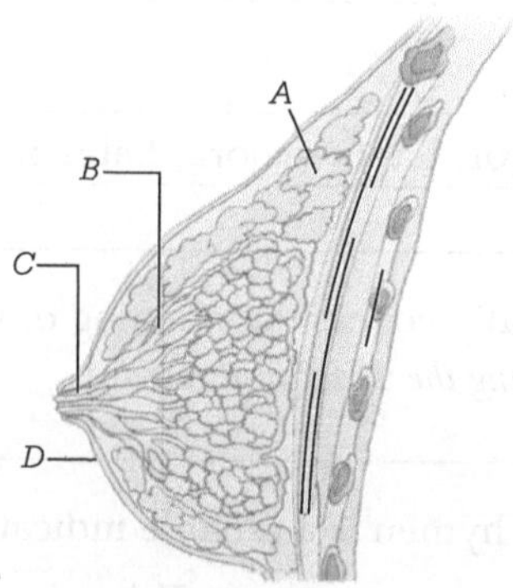

A. ______________ B. ______________ C. ______________ D. ______________

40. Match the following columns.

	Column I (Parts of mammary gland)		Column II (Characteristics)
A.	Mammary duct	1.	Join to form mammary ampulla
B.	Lactiferous duct	2.	Secretes milk
C.	Mammary tubules	3.	Duct through which milk is sucked out
D.	Cells of alveoli	4.	Alveoli open into it

A. ______________ B. ______________ C. ______________ D. ______________

Topic 3 : Gametogenesis

41. Number of chromosomes present in spermatogonium is ______________

42. Which cells come earliest in the sequence of sperm production?

Spermatozoa, Spermatocyte, Spermatid, Spermatogonia

____

43. Spermatogenesis is initiated in the male testis with the beginning of ______________ (puberty/foetal stage).

44. ______________ (Mitosis or Meiosis) division is involved in the formation of secondary spermatocytes from primary spermatocytes.

45. Sperms are released from the <u>Sertoli cells</u> by a process called spermiation. (*Rewrite the statement by changing the underlined word.*)

____

46. Number of chromosomes present in secondary spermatocyte is ______________

47. During spermatogenesis, which cells are the first to contain haploid number of chromosomes? ______________

48. Find out spermatid and Sertoli cell in given below diagram.

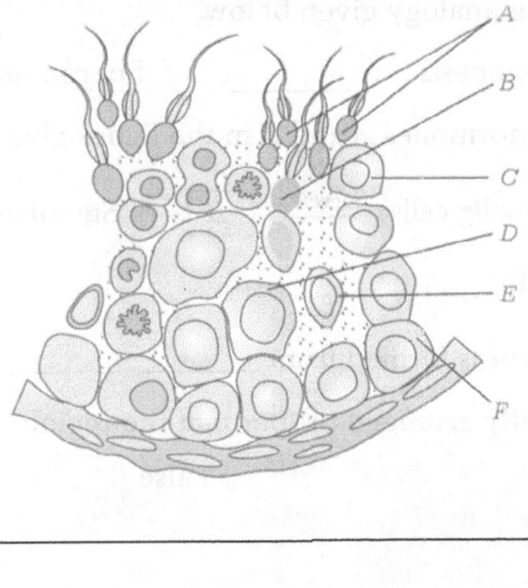

____

49. Name the process by which spermatids transformed into spermatozoa.

____

50. Complete the following flowchart of sperm formation.

Spermatogonia ⟶ *A* ⟶ Spermatid ⟶ *B*

A. ______________ B.______________

51. Increase in the GnRH hormone secretion leads to spermatogenesis.

True ☐ False ☐

52. Name two hormones stimulated by GnRH during spermatogenesis. ______________

53. Identify the processes *A, B, C* and *D* in the schematic representation of spermatogenesis given below.

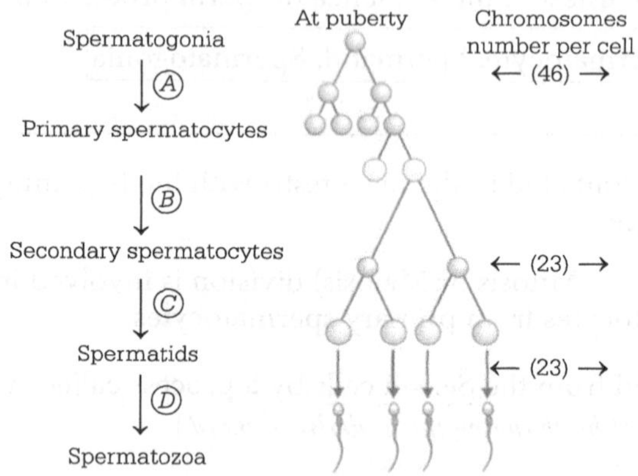

A. ______________ B.______________ C.______________ D.______________

54. Complete the following analogy given below.

Luteinizing hormone : Leydig cells : : ______________ : Sertoli cells.

55. Complete the following analogy given below.

Androgens : Spermatogenesis : : ___________: Spermiogenesis

56. Give the name of two hormones *A* and *B* in the figure given below.

A —Acts on⟶ Leydig cells —Secrete⟶ *B* ⟶ Stimulate spermatogenesis

A. ______________ B.______________

57. Acrosome cone in sperm is formed from ______________ (Golgi bodies/mitochondria).

58. Mitochondria are spirally arranged in sperm at the region of middle piece.

True ☐ False ☐

59. For normal fertility, ____________ % of sperm ejaculate must exhibit vigorous motility and _________ % sperms must have normal shape and size.

60. In the given figure of structure of sperm , identify the incorrect labellings.

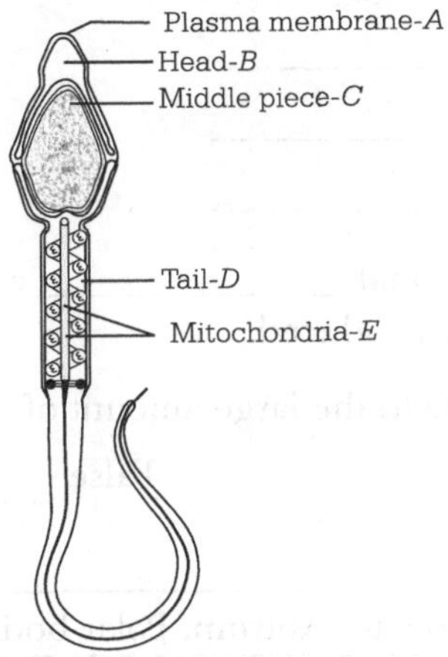

____

61. Choose the incorrect pair.

(i) Acrosome – Enzyme filled cap structure on sperm head
(ii) Middle piece – Possesses numerous mitochondria
(iii) Sperm tail – Facilitates sperm motility essential for fertilisation
(iv) Semen – Contains only sperms

____

62. Secretions of epididymis, vas deferens, seminal vesicle and prostate are essential for ______________ (acrosome reaction/maturation and motility of sperms).

63. The function of male sex accessory ducts and glands is maintained by androgens.

True ☐ False ☐

64. Gamete mother cells (oogonia) start division and get arrested at ______________ stage of meiotic division as ____________________________ oocytes.

65. At puberty, only 60,000-80,000 primary follicles are left in each ovary.

True ☐ False ☐

66. In the given figure, correctly label the structures *A, B, C* and *D*.

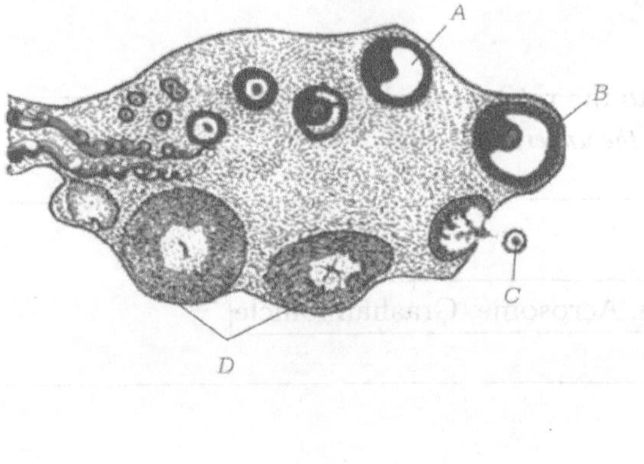

A. ______________________

B. ______________________

C. ______________________

D. ______________________

67. Name the fluid-filled cavity, which is a characteristic of a tertiary follicle.

____

68. Write the ploidy of the following cells.

    (i) Secondary oocyte_______________

    (ii) Polar body _______________

    (iii) Primary oocyte _______________

69. During oogenesis, one functional _______________ and three tiny _______________ are produced.

70. Large size of the ovum is due to the large amount of reserve food.

    True ☐ False ☐

71. Find the odd one out.

Cowper's gland, Primary oocyte, Antrum, Polar bodies

    ____

72. Structure consisting of ova with their surrounding tissues is called _______________

73. Name the membrane surrounding the secondary oocyte.

    ____

74. Rupture of zona pellucida releases the secondary oocyte from the ovary.
    (*Rewrite the statement by changing the underlined word.*)

    ____

75. Identify the labelled parts as *A, B, C, D* and *E* from the options given below.

Theca interna, Theca externa, Ovum, Antrum, Membrana granulosa

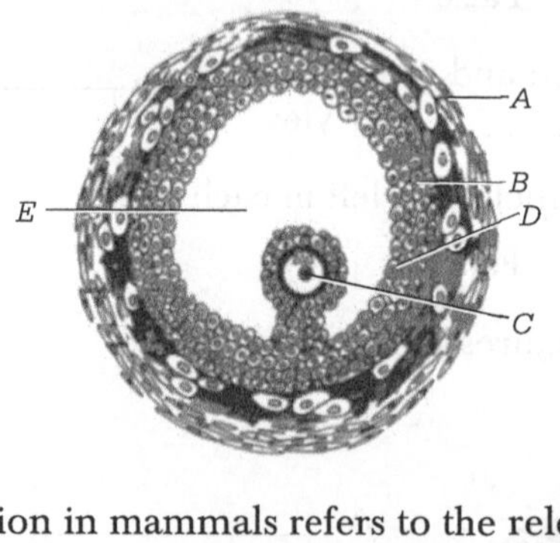

A. _______________

B. _______________

C. _______________

D. _______________

E. _______________

76. Ovulation in mammals refers to the release of primary oocyte in ovary.
    (*Rewrite the statement by changing the underlined word.*)

    ____

77. Find the odd one out.

Endometrium, Corpus luteum, Acrosome, Graafian follicle

    ____

78. Mitotic differentiation of oogonia occurs during fetal life.

True ☐ False ☐

79. Select the correct pair (s) among the following.
(i) Secondary oocyte – Haploid
(ii) Primary oocyte – Haploid
(iii) Ovum – Diploid
(iv) Primary spermatocyte – Diploid

____

80. Choose the incorrect pair.
(i) Antrum – Fluid-filled cavity in primary follicle
(ii) Tertiary follicle – Primary oocyte completes its Ist meiotic division inside it
(iii) Secondary oocyte – Haploid cell formed after Ist meiotic division
(iv) Graafian follicle – Mature tertiary follicle which ruptures during ovulation

____

81. Match the following columns.

	Column I		Column II
A.	Ovum	1.	Uterus
B.	Endometrium	2.	Oogenesis
C.	Polar body	3.	Secondary oocyte
D.	Clitoris	4.	Rudimentary erectile tissue

A. ____________ B. ____________ C. ____________ D. ____________

82. The number of chromosomes in a mature gamete is reduced to half during the formation of first polar body.

True ☐ False ☐

83. Complete the following flowchart of oogenesis.

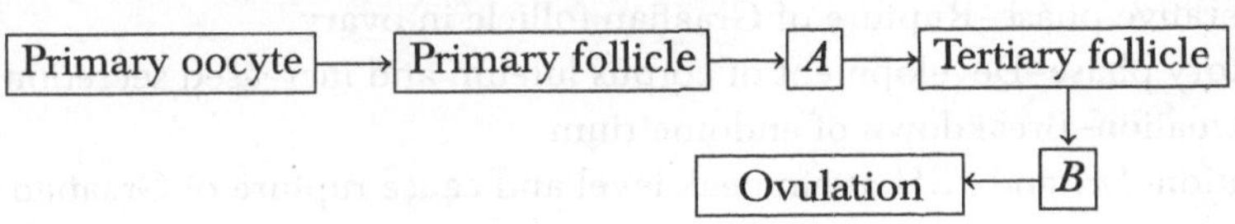

A.____________ B.____________

Topic 4 : Menstrual Cycle

84. Write the scientific name for the first menstruation at puberty.

____

85. The reproductive cycle repeated at an interval of about____________ days, in human female is called menstrual cycle.

86. When does ovulation occur in a healthy menstruating female?

____

87. Endometrium, epithelial glands and connective tissue of uterus are broken due to the lack of progesterone, thus resulting in menstrual flow.

True ☐ False ☐

88. What happens in uterus during the follicular phase of menstrual cycle?

____

89. During follicular phase, oestrogen is secreted by _______________ (pituitary/growing follicles).

90. Rapid secretion of LH or LH surge in ovulatory phase causes

Rupturing of Graafian follicle, Release of ova, Ovulation

____

91. The corpus luteum secretes large amount of FSH during luteal phase. (*Rewrite the statement by changing the underlined word.*)

____

92. No new follicles develop in the luteal phase of the menstrual cycle because

____

93. If the mammalian ovum fails to get fertilised, which one of the following is likely to happen?

Corpus luteum will degenerate, Oestrogen secretion further decreases, Primary follicle starts developing, Progesterone secretion rapidly increases

____

94. Which of them is not a correct match?
(i) Proliferative phase–Rupture of Graafian follicle in ovary
(ii) Secretory phase–Development of corpus luteum and increased secretion of progesterone
(iii) Menstruation–Breakdown of endometrium
(iv) Ovulation–LH and FSH attain peak level and cause rupture of Graafian follicle

____

95. Write the correct sequence of secretions of hormones given in box from beginning of menstrual cycle to the end.

Progesterone, FSH, Oestrogen

____

96. The following graph of relative concentrations of the four hormones present in the blood plasma of a woman during her menstrual cycle. Identify the hormones *A*, *B*, *C* and *D*.

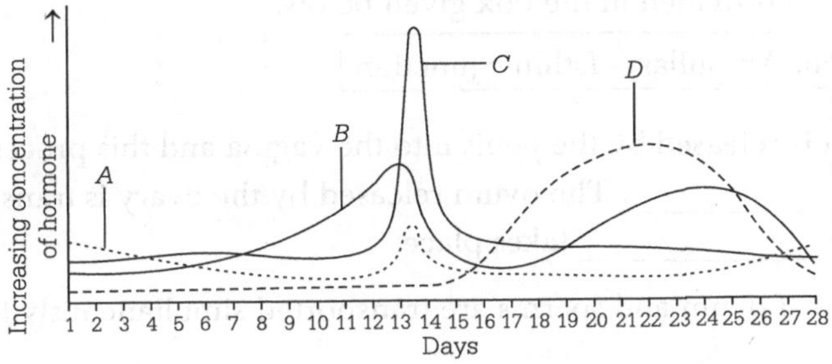

A. __________________

B. __________________

C. __________________

D. __________________

97. The diagram shows the changes that take place in the endometrium during a normal menstruation. Identify *A* and *B*.

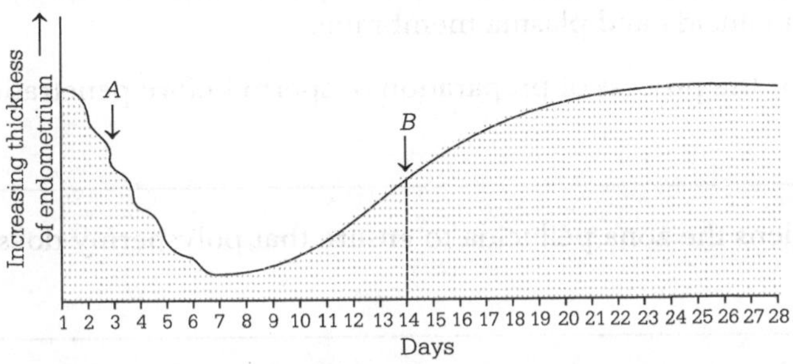

A. __________________

B. __________________

98. Some important events in the human female reproductive cycle are given below. Arrange the events in proper sequence.

I. Secretion of FSH.
II. Growth of corpus luteum.
III. Growth of the follicle and oogenesis.
IV. Ovulation.
V. Sudden increase in the levels of LH.

____

99. Match the following columns.

	Column I		Column II
A.	Lack of the menstruation	1.	LH surge
B.	Secretory phase	2.	Corpus luteum development and increased progesterone
C.	Ovulation	3.	Cessation of menstrual cycle
D.	Menopause	4.	Indicating of pregnancy

A. __________ B. __________ C. __________ D. __________

Topic 5 : Fertilisation and Implantation

100. Fill in the blanks with words provided in the box given below.

Fertilisation, Insemination, Ampullary- Isthmic junction

During copulation, semen is released by the penis into the vagina and this process is called ______________________ . The ovum released by the ovary is transported to the ____________ where ____________ takes place.

101. Fertilisation can only occur if ovum and sperms are transported simultaneously to ampullary region.

True ☐ False ☐

102. The secretions of ______________________ help the sperm enter into the cytoplasm of ovum through the zona pellucida and plasma membrane.

103. Write the scientific name for the process of preparation of sperm before penetration of ovum.

____

104. A reaction of granules hardens the zona pellucida to ensure that polyspermy does not occur. Name this reaction.

____

105. Corona radiata ensures that only one sperm can fertilise the ovum. (*Rewrite the statement by changing the underlined word*).

____

106. In females, meiosis-II is not completed until fertilisation.

True ☐ False ☐

107. Secondary maturation of oocyte occurs in the ovary. (*Rewrite the statement by changing the underlined word.*)

____

108. Arrange the events that take place during fertilisation given below in a proper sequence.

Cortical reaction, Insemination, Karyogamy, Sperm entry

____

109. Sex of the foetus is decided during fertilisation by male gamete.

True ☐ False ☐

110. Cleavage differs from mitosis in lacking ______________ (synthetic/growth) phase.

111. Cleavage forms 2-4 to 8-16 cells. These cells are called______________________

112. Name the structure formed from zygote, with 8-16 cells after successive mitotic division. ______________________

113. After blastocyst formation, the ____________ gets attached to the endometrium and the ____________ gets differentiated as the embryo.

114. Trophoblast of blastocyst attaches to the myometrium.
(*Rewrite the statement by changing the underlined word.*)

____

115. During implantation, blastocyst gets embedded in the endometrium of uterus.

True ☐ False ☐

116. The given diagram refers to ovum surrounded by few sperms. Identify the labelled parts as *A*, *B* and *C* in the diagram.

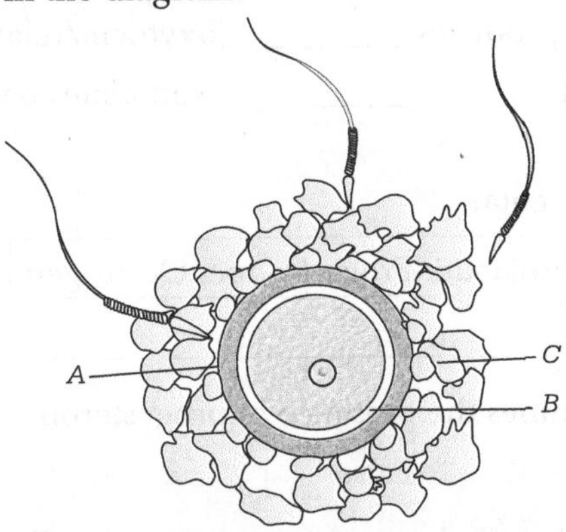

A. ____________________

B. ____________________

C. ____________________

117. Refer to the given diagram and identify the structures *A* and *B*.

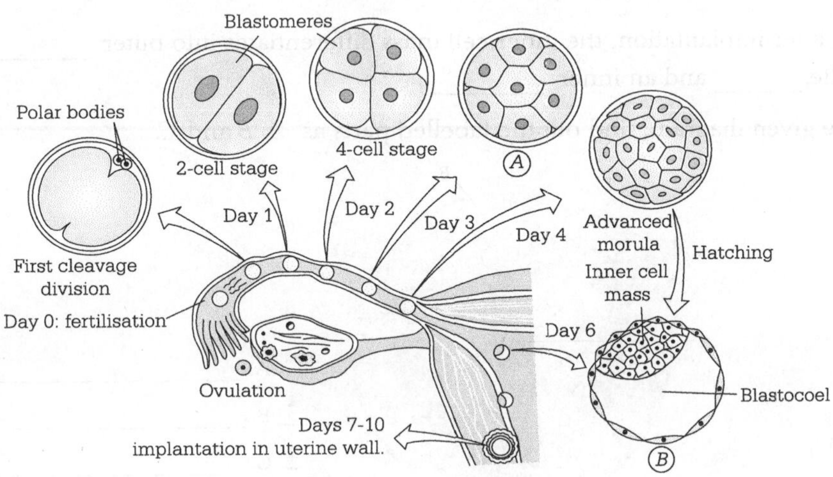

A. ____________________ B. ____________________

118. Complete the flowchart according to the sequence of events taking place in fertilisation.

Cleavage → *A* → Morula
Morula ↓ Blastula
Blastula → Trophoblast
Blastula ↓ *B* ↓ Embryo ↓ Implantation

A. ____________________

B. ____________________

Topic 6 : Pregnancy and Embryonic Development

119. Give the scientific name for the finger-like projections on the trophoblast. ___________

120. ________________________ is the structure which connects foetus to the maternal body.

121. Placenta is connected to the embryo through cervical canal.

True ☐ False ☐

122. Name any three hormones produced by placenta.________________________

123. In later stages of pregnancy, ovary produces ________ (oxytocin/relaxin).

124. In a pregnant woman, high level of ________________ stimulates oestrogen and progesterone synthesis.

125. Find the odd one out related to placenta.

Human chorionic gonadotropin, Prolactin, Progesterone, Oestrogen

____

126. Urine test during pregnancy determines the presence of progesterone.

True ☐ False ☐

127. Inner cell mass contains certain cells called __________________ cells which have the potency to give rise to all the tissues and organs.

128. Soon after implantation, the inner cell mass differentiates into outer _________, middle ______ and an inner ________ .

129. In the given diagram, find out the labelled parts as *A*, *B* and *C*.

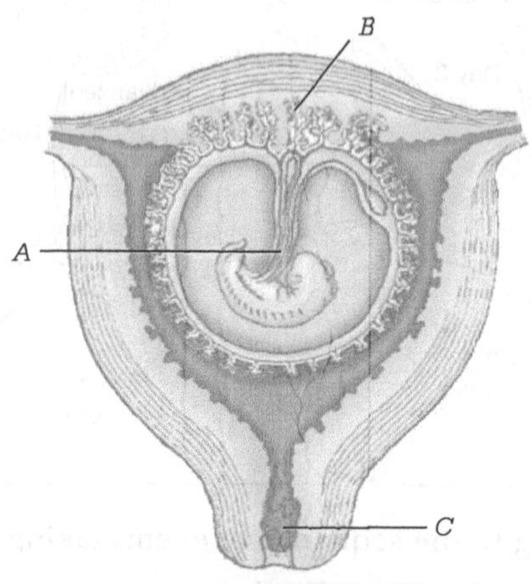

A. ____________________

B. ____________________

C. ____________________

130. During embryonic development, differentiation of <u>organs</u> occurs first.
(*Rewrite the statement by changing the underlined word.*)

____

131. Name the embryonic layers from which following structures develop.

(i) Epidermis ____________________

(ii) Teeth enamal ____________________

(iii) Muscles ____________________

132. The amnion of mammalian embryo is derived from

____

133. Choose the incorrect pair.

	Organ		Time of development in foetus
(i)	Heart	–	After one month of pregnancy
(ii)	Body covered with fine hair	–	By the end of 24 weeks
(iii)	External genital organs	–	By the end of 26 weeks (second trimester)
(iv)	First movement of foetus	–	During the fifth month

____

134. The foetus develops limbs and digits by the end of fourth month of pregnancy.
(*Rewrite the statement by changing the underlined word.*)

____

135. By the end of ____________ trimester, most of the major organ systems are developed.

Topic 7 : Parturition and Lactation

136. Write the scientific name for delivery of developed foetus. ____________

137. Fll in the blanks with words provided in the box given below.

Placenta, Parturition, Oxytocin, Prolactin, Amnion

The signals for ____________ originate from the fully developed foetus and ____________, which induces mild uterine contractions called foetal-ejection reflex. This triggers the release of ____________.

138. Name the hormone responsible for strong uterine contractions during parturition.

____

139. The process of parturition is induced by a complex muscular mechanism.
(*Rewrite the statement by changing the underlined word.*)

____

140. Find the odd one out.

Cortisol, Oestrogen, Oxytocin, Follicle stimulating hormone

____

141. Oxytocin is released from the ____________ (anterior/posterior) lobe of the pituitary gland.

____

142. Write the scientific name for the milk produced during the initial few days of lactation.

____

143. Colostrum, released from the mammary gland is rich in several antibodies.

True ☐ False ☐

144. Match the following columns.

Column I	Column II
A. Gestation	1. Delivery of baby from uterus
B. Lactation	2. Attachment of embryo to endometrium
C. Parturition	3. Duration between pregnancy and birth
D. Implantation	4. Production of milk after delivery

A. ______________ B. ______________ C. ______________ D. ______________

145. Breastfeeding immediately after delivery is not recommended by doctors.

True ☐ False ☐

146. Arrange the following events in the sequence of process of reproduction in humans.

Parturition, Gametogenesis, Fertilisation, Implantation, Lactation

____

Answers

1. Scrotum
2. Scrotum
3. (i) 4-5 cm, (ii) 2-3 cm
   (iii) 250, (iv) 1-3
4. Spermatogonia
5. Sertoli cells
6. Spermatogonia
7. Interstitial or Leydig cells
8. True
9. Rete testis, vas deferens, urethral meatus
10. (i) Leydig cells secrete androgens.
11. (iii) is correct.
12. Urethra
13. *A*–Urinary bladder, *B*–Urethra, *C*–Vas deferens, *D*–Bulbourethral gland
14. Seminal vesicle is an accessory gland.
15. *A*–Sertoli cells, *B*–Spermatogonia, *C*–Interstitial cells, *D*–Spermatozoa
16. Androgens
17. Statement (ii) is incorrect because testosterone plays a stimulatory role in the formation of spermatozoa.
18. True
19. Seminiferous tubule is odd one out. Rest are parts of male sex accessory ducts.
20. Fructose, calcium, enzymes.
21. Bulbourethral gland
22. Foreskin or prepuce
23. Ligaments
24. True
25. (ii) is incorrect because ovaries produce several steroid hormones. LH and FSH are secreted by pituitary.
26. Female accessory ducts
27. *A*–Uterus, *B*–Urinary bladder, *C*–Urethra, *D*–Cervix, *E*–Vagina
28. *A*-Endometrium
    C-Perimetrium
29. (i) Fimbriae, (ii) Isthmus, (iii) Cervix
30. Endometrium
31. Pectoralis major is odd one out. Rest are the structures of female external genitalia.
32. Hymen
33. False. Hymen can also be broken by sudden fall or jolt, active participation in cycling, horse-back riding, etc.
34. (iv) is incorrect because myometrium exhibits strong contractions during delivery.
35. I-Fallopian tube, II-Ejaculatory duct, III-Acrosome, IV-Testis
36. A-2, B-1 C-3, D-4
37. *A* – Mammary duct
    B – Lactiferous duct
38. Glandular
39. *A*–Fat, *B*–Mammary duct, *C*–Lactiferous duct, *D*–Areola
40. A-1, B-3, C-4, D-2
41. 46
42. Spermatogonia
43. Puberty
44. Meiosis
45. Seminiferous tubules
46. 23
47. Secondary spermatocyte
48. *B*–Spermatid and *E*–Sertoli cell
49. Spermiogenesis
50. *A* - Spermatocyte
    B - Sperms
51. True
52. Luteinizing Hormone (LH) and Follicle Stimulating Hormone (FSH).
53. *A*-Mitosis and differentiation
    B-1st meiotic division
    C-2nd meiotic division
    D-Differentiation
54. Follicle Stimulating Hormone (FSH)
55. Follicle Stimulating Hormone (FSH)
56. LH and androgens
57. Golgi bodies
58. True
59. 40, 60
60. Incorrect labellings are *B*-Acrosome, *C*-Head, *D*-Middle piece
61. (iv) is incorrect because semen contain sperms, seminal fluid, secretions of prostate glands, etc.
62. Maturation and motility of sperms.
63. True
64. Prophase-I, primary
65. True
66. *A*-Tertiary follicles showing antrum.
    B-Graafian follicle
    C-Secondary oocyte
    D-Corpus luteum
67. Antrum
68. (i) Haploid,
    (ii) Haploid,
    (iii) Diploid
69. Ovum, polar bodies
70. True
71. Cowper's gland is odd one out. Rest are structures related to oogenesis.
72. Graafian follicle
73. Zona pellucida
74. Graafian follicle
75. *A* - Theca externa
    B - Theca interna
    C - Ovum
    D - Membrana granulosa
    E - Antrum
76. Secondary oocyte.
77. Acrosome is odd one out. Rest are structures related to female reproductive system.
78. True
79. (iv) is correct. Secondary oocyte and ovum are haploid and primary oocyte is diploid.
80. (i) is incorrect because antrum is fluid filled cavity in tertiary follicle.

81. A-3, B-1, C-2, D-4

82. True

83. *A* - Secondary follicle
B - Secondary oocyte

84. Menarche

85. 28 / 29

86. Middle of cycle, i.e. 14-16 days

87. True

88. Endometrium proliferation

89. Growing follicles

90. LH surge causes rupture of Graafian follicle due to which ova is released and this phase is called ovulation.

91. Progesterone

92. Follicles do not remain in the ovary after ovulation.

93. Corpus luteum will degenerate.

94. (i) is incorrect because during proliferative phase, Graafian follicle is formed in ovary.

95. FSH, oestrogen, progesterone

96. *A*-FSH, *B*-Oestrogen, *C*-LH, *D*-Progesterone

97. A-Menstruation, B-Ovulation

98. I → III → V → IV → II

99. A-4, B-2, C-1, D-3

100. Insemination, ampullary - isthmic junction, fertilisation

101. True

102. Acrosome

103. Capacitation

104. Cortical reaction

105. Zona pellucida

106. True

107. Fallopian tube

108. Insemination → Sperm entry → Cortical reaction → Karyogamy

109. True

110. Growth

111. Blastomeres

112. Morula

113. Trophoblast, Inner cell mass

114. Endometrium

115. True

116. *A*–Zona pellucida, *B*–Perivitelline space, *C*–Corona radiata

117. *A*-Morula , *B*-Blastocyst

118. *A*-Blastula
B-Inner cell mass

119. Chorionic villi

120. Placenta

121. False. Placenta is connected to the embryo through umbilical cord.

122. hCG, hpL, Oestrogen

123. Relaxin

124. Human Chorionic Gonadotropin (hCG).

125. Prolactin is odd one out. Rest are the hormones produced by placenta.

126. False. Urine test determines the presence of Human Chorionic Gonadotropin (hCG).

127. Stem

128. Ectoderm, mesoderm, endoderm

129. *A*–Umbilical cord with its vessels, *B*–Placental villi, *C*–Plug of mucus in cervix

130. Cells

131. (i) Ectoderm, (ii) Endoderm, (iii) Mesoderm

132. Ectoderm and mesoderm

133. (iii) is incorrect because external genitilia develop by the end of 12 weeks.

134. Second

135. First

136. Parturition

137. Parturition, placenta, oxytocin

138. Oxytocin

139. Neuroendocrine

140. Follicle Stimulating Hormone (FSH) is odd one out. Rest are directly involved in parturition.

141. Posterior

142. Colostrum

143. True

144. A-3, B-4, C-1, D-2

145. False. Breastfeeding immediately after delivery is always recommended by the doctors as the milk produced during initial few days of lactation is rich in antibodies, which will help develop resistance in the newborn babies.

146. Gametogenesis → Fertilisation → Implantation → Parturition → Lactation

CHAPTER FOUR

REPRODUCTIVE HEALTH

Topic 1 : Reproductive Health : Problems and Strategies

1. According to WHO, reproductive health means a total well-being in all aspects of reproduction, i.e. physical, emotional, behavioural and social.

   True ☐ False ☐

2. India was the ______________ (first/fifth) country in the world to have launched a National Programme for Family Planning in 1951.

3. A reproductively healthy society do not have people with
   (i) physically normal reproductive organs
   (ii) functionally anomalous reproductive organs

   ____

4. Write the full form of
   (i) RCH ____
   (ii) CDRI ____

5. RCH programme was launched in April, 2005.

   True ☐ False ☐

6. Proper information about reproductive organs, adolescence and related changes would help people in ______________ age group to lead a reproductively healthy life.

7. Complete the sentence with the words given in the box below.

Reproductive, Physioemotional, Informational, Infrastructural, Economical

Implementation of various action plans to attain ______________ health requires strong ______________ facilities, professional expertise and material support.

8. Give the scientific name for the foetal sex-determination test. ____________

9. Amniocentesis is based on the ____________ pattern in the ____________ fluid surrounding the developing embryo.

10. Statutory ban on amniocentesis for sex-determination is to check increasing female foeticide.

True ☐ False ☐

Topic 2 : Population Stabilisation and Birth Control

11. What proportion of world's population lives in India?

____

12. The world population in 2011 was ____________ (2.8 billion/7.2 billion).

13. According to the 2011 census report, the population growth rate was ______/ 1000/year.

14. Write the full forms of the following.

(i) MMR ____________

(ii) IMR ____________

15. Complete the following diagram.

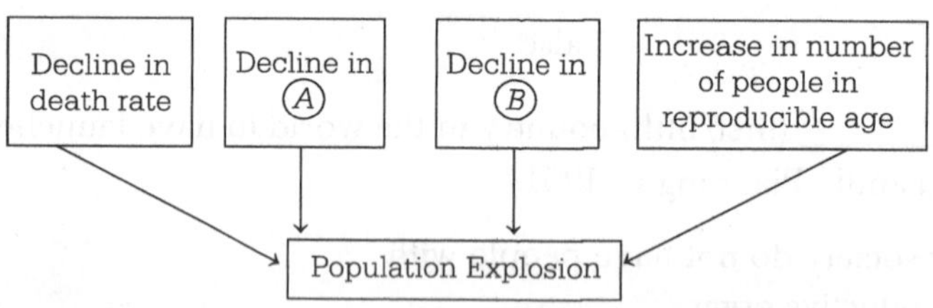

A. ____________

B. ____________

16. Select the main factors from the box given below affecting the population growth.

Industrialisation, Increase in healthcare facilities, Increase in birth rate, Decrease in birth rate, Decrease in death rate

____

17. Decreasing the death rate is the main goal to control population.

True ☐ False ☐

18. The use of prophylactic vaccinations and antibiotics is directly responsible for the current rapid growth in population.

True ☐ False ☐

19. Birth rate is the number of live births per ____________ (100/1000) people per year.

20. The rate at which newborn individuals are joining the population by reproduction is known as <u>fertility</u>.
(*Rewrite the statement by changing the underlined word.*)

____

21. Marriageable age of female is __________ years and males is __________ years in India.

22. Give the scientific name of the deliberate use of artificial methods or techniques to prevent pregnancy. ______________________________

23. Write any two requirements for an ideal contraceptive.
(i) ______________________________ (ii) ______________________________

24. Natural methods work on the principle of avoiding chances of ovum and sperm meeting.
True ☐ False ☐

25. Give the biological name for the contraceptive method in which couples abstain from coitus.
____

26. Periodic abstinence is also known as rhythm method or calendar method.
True ☐ False ☐

27. Fill in the blanks using the words given in the box below.

Abstain, Ovulation, Coitus, Reproductive, Menstrual, Copulation

Periodic abstinence is one such method in which the couple avoid or ____________ from ____________ from 10-17 days of the ____________ cycle when ____________ could be expected.

28. Name the method in which the male partner withdraws his penis from vagina before ejaculation. ______________________________

29. Fertilisation is prevented in withdrawal method by avoiding ______________ (insemination/ovulation).

30. The effectiveness of coitus interruption as contraception is least among the other natural methods.
True ☐ False ☐

31. Write the scientific name of a natural method of contraception due to intense lactation period. ______________________________

32. ______________ (Ovulation/Insemination) does not occur during the period of intense lactation following parturition.

33. Conception is almost nil for a breastfeeding mother.
True ☐ False ☐

34. Lactational amenorrhea is effective only upto ______________ months following parturition.

35. Chances of failure of natural methods is <u>low</u>, with no side effects.
(*Rewrite the statement by changing the underlined word.*)
____

36. Barrier method involves preventing physical meeting of ____________ and ____________ with the help of barriers.

37. Name the barriers made up of thin rubber/latex sheath used for the physical prevention of gametes meeting. ____________

38. Condoms can be used to cover both vagina and cervix in females.

True ☐ False ☐

39. Write the name of the popular brand of condom for the male in India.

____

40. Use of condoms increased over the years due to its additional benefit of protecting the user from contracting ____________ and ____________.

41. Find the odd one out regarding condoms.

User friendly, Easily disposable, Self-insertable, Reusable

____

42. Fill in the blanks with the words given below in the box.

Diaphragms, Cervical caps, Vaults, Condoms, Copper-T

____________, ____________ and ____________ are the contraception exclusively for females and are reusable, which block the entry of sperms.

43. Spermicidal creams decrease the contraceptive efficiency.

True ☐ False ☐

44. Write the full form of IUD. ____________

45. <u>Condoms</u> should be inserted with help of a doctor or expert nurses.
(*Rewrite the statement by changing the underlined word.*)

____

46. IUDs are placed in ____________ (uterus/vagina).

47. Which of the following pairs are incorrectly matched.
(i) Lippe's loop – Non-medicated IUD.
(ii) Progestasert – Copper releasing IUD.
(iii) LNG-20 – Hormone releasing IUD.

____

48. Give a common term for the following.
CU-T, Cu-7, Multiload-375 ____________

49. IUDs increase ____________ (phagocytosis/pinocytosis) of sperms within the uterus.

50. Cu ions ____________ sperm motility and fertilising capacity of sperms.

51. IUDs make the uterus suitable for implantation.

True ☐ False ☐

52. Match the following columns.

	Column I (Contraception methods)		Column II (Features)
A.	Condoms	1.	Increase sperm phagocytosis in cervix
B.	Diaphragms	2.	User friendly latex sheath
C.	IUDs	3.	High chances of failure
D.	Lactational amenorrhea	4.	Self inserted and reusable

A. ______________ B. ______________ C. ______________ D. ______________

53. Oral pills used as contraceptives, contain small doses of ____________________ or combination of ____________________ .

54. Oral administration has to be taken daily for a period of ____________ starting within first five days of menstrual cycle.

55. Pills inhibit ______________ and implantation. They also alter the quality of cervical mucus to prevent entry of ______________ .

56. Pills are very effective with ______________ (greater/lesser) side effects and are well accepted by ______________ (females/males).

57. ____________________ is a 'once a week' pill containing non-steroidal prepration, centchroman.

58. Progestogen used as implants under skin have longer effective periods.

True ☐ False ☐

59. Pills and implants prevent the pituitary from secreting ______________ and __________ so that ovulation does not occur.

60. Emergency pills are taken within ____________ hours after coitus to avoid possible pregnancy.

61. Find the odd one out.

Vasectomy, Tubectomy, Copper-T

____

62. Complete the following analogy given below.

Males : Females : : ________________ : Tubectomy

63. ______________ (Epididymis/Vas deferens) is tied and cut in the process of vasectomy.

64. After vasectomy semen does not contain sperms.

True ☐ False ☐

65. Identify *A, B, C* and *D*.

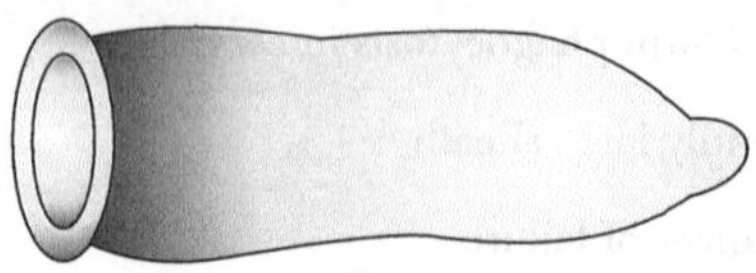

A

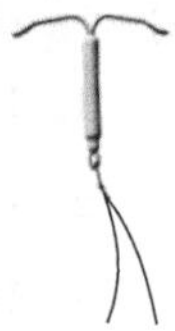

B

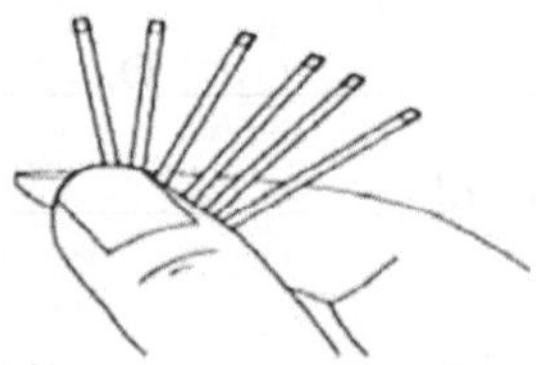

C

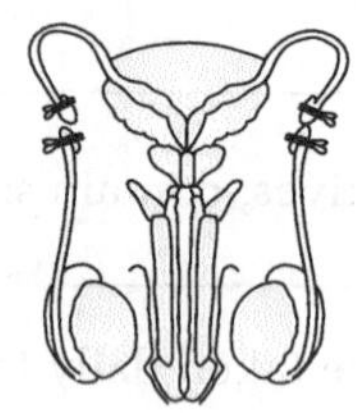

D

A. ____________

B. ____________

C. ____________

D. ____________

66. Name the process and the reproductive structure involved in the process.

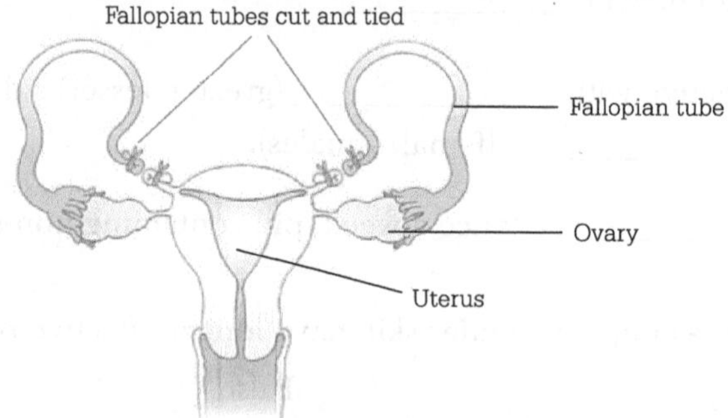

____

67. After tubectomy, ovum would not be found in
(i) uterus
(ii) ovaries

____

68. Reversibility is high for sterilisation techniques.

True ☐ False ☐

69. Find the odd one out.

Nausea, Abdominal pain, Menstrual bleeding, Fever

____

Topic 3 : Medical Termination of Pregnancy (MTP)

70. Write the full form of MTP. ____________________

71. MTP is the <u>involuntary</u> termination of pregnancy before full term.
(*Rewrite the statement by changing the underlined word.*)

____

72. Answer the following questions in numbers only.

(i) Number of MTPs performed per year across the world.____________________

(ii) The time period (in weeks) upto which MTP is relatively safe. ____________________

(iii) The year in which government of India legalised MTP. ____________________

(iv) The year in which MTP act was enacted by government of India. ____________________

(v) Number of opinions required by medical practitioners if MTP has to be performed after 12 weeks. ____________________

73. MTPs are not essential and legal in which of the following cases?
(i) To get rid of unwanted pregnancies.
(ii) To check illegal female foeticide.
(iii) When continuation of pregnancy could be harmful for mother or foetus.

____

74. Write dangerous trends associated with MTP besides female foeticide.

____

Topic 4 : Sexually Transmitted Diseases (STDs)

75. Give the scientific term used for diseases or infections that are transmitted through sexual intercourse. ____________________

76. Write the other names of STDs. ____________________

77. Find the odd one out.

Genital herpes, Syphilis, Trichomoniasis, Amoebiasis

____

78. Through which of the following modes given below the transmission of STDs is not possible.
(i) Transfusion of blood from Rh negative donor.
(ii) Sharing of infected needles.
(iii) Sexual relationship with multiple partners.

____

79. Find the odd one out.

Genital herpes, HIV, Hepatitis-B, Chlamydiasis

____

80. Absence of significant symptoms in the early stages of infection and social stigmas attached to the STDs deter the person from getting proper treatment.

True ☐ False ☐

81. Write the most vulnerable age groups to STDs. ______________________

82. Identify *A* and *B*.

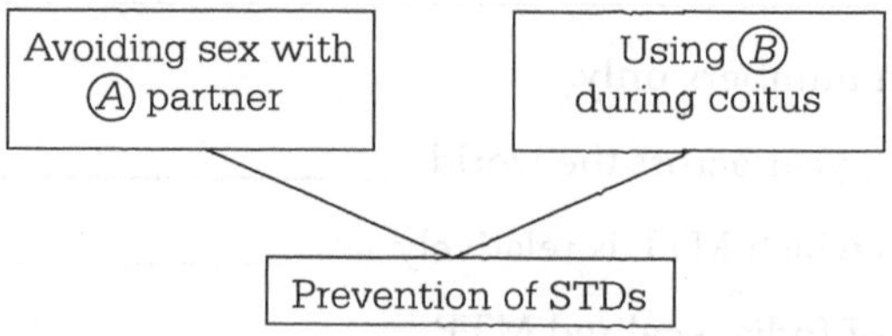

A. ______________________

B. ______________________

83. Nematodes can cause serious Sexually Transmitted Diseases (STDs).

True ☐ False ☐

84. Mention any two complications of STDs in reproductively active women.

____

85. Name the microbe causing syphilis. ______________________

86. AIDS is diagnosed through ELISA test.

True ☐ False ☐

87. Name the disease diagnosed by the identification of antibodies against antigen.

____

88. Write the full form of ELISA. ______________________

89. Match the following columns.

Column I (Diseases)	Column II (Causative organisms)
A. Gonorrhea	1. HIV
B. Syphilis	2. Human papilloma virus
C. Genital warts	3. *Neisseria*
D. AIDS	4. *Treponema*

A. ____________ B. ____________ C. ____________ D. ____________

Topic 5 : Infertility

90. ________________ is the inability of couples in the reproducible age group to produce offspring inspite of unprotected sexual to habitation.

91. Give the common term for the special assisted techniques to have children.

____

92. IVF stands for *In Vivo* Fertilisation.

True ☐ False ☐

93. Complete the statements by choosing the right words from the box given below.

Fallopian tube, Uterus, ZIFT, IUT

The zygote or embryo, upto eight blastomeres are transferred to __________ . The process is called __________ .

94. Complete the following analogy given below.

8 blastomeres : __________ :: More than 8 blastomeres : ________________

95. Find the odd one out.

IVF, PID, ZIFT, GIFT

____

96. Write the full form of GIFT. ____

97. ________________ is a specialised procedure to form an embryo in which a sperm is directly injected into the ovum.

98. IUI is a technique in which semen collected from the husband or a healthy donor is artificially introduced into the <u>cervix</u>.
(*Rewrite the statement by changing the underlined word.*)

____

99. Match the following columns.

Column I	Column II
A. IVF	1. *In vitro* fertilisation, embryo transferred into Fallopian tube.
B. ZIFT	2. *In vitro* fertilisation, embryo implantation into uterus.
C. GIFT	3. Pre-natal diagnosis of foetus for genetic disorder.
D. AFT	4. *In vivo* fertilisation, sperm and ova transferred in Fallopian tube.

A. ____________ B. ____________ C. ____________ D. ____________

100. Adoption is considered as one of the best methods for couples looking for parenthood.

True ☐ False ☐

Answers

1. True
2. First
3. (ii), In a reproductive healthy society people do not have functionally anomalous reproductive organs.
4. (i) Reproductive and Child Healthcare
   (ii) Central Drug Research Institute
5. True
6. Adolescent
7. Reproductive, infrastructural
8. Amniocentesis
9. Chromosomal, amniotic
10. True
11. 17.7%
12. 7.2 billion
13. 20
14. (i) Maternal Mortality Rate
    (ii) Infant Mortality Rate
15. *A*-Maternal Mortality Rate (MMR)
    B-Infant Mortality Rate (IMR)
16. Increase in healthcare facilities,
    Increase in birth rate, Decrease in death rate.
17. False. Decreasing the birth rate is the main goal to control population.
18. True
19. 1000
20. Natality
21. 18, 21
22. Contraception
23. (i) Easy availability
    (ii) Least or no side effect
24. True
25. Periodic abstinence
26. True
27. Abstain, coitus, menstrual, ovulation
28. Coitus interruptus or withdrawl
29. Insemination
30. True
31. Lactational amenorrhea
32. Ovulation
33. True
34. Six
35. High
36. Sperm, ova
37. Condom
38. True
39. Nirodh
40. STD, AIDS
41. Reusable is odd one out. Rest of the options are the advantages of using condom as contraceptive.
42. Diaphragms, cervical caps, vaults
43. False. Spermicidal creams increase the contraceptive efficiency.
44. Intra Uterine Device
45. IUDs
46. Uterus
47. (ii) is incorrect because progestasert is hormone releasing IUD.
48. Copper-releasing IUDs
49. Phagocytosis
50. Suppress
51. False. IUDs make the uterus unsuitable for implantation.
52. A-2, B-4, C-1, D-3
53. Progestogens, progestogen oestrogen
54. 21 days
55. Ovulation, sperms
56. Lesser, females
57. Saheli
58. True
59. FSH , LH
60. 72
61. Copper T is odd one out. Rest two are surgical methods of sterilisation.
62. Vasectomy
63. Vas deferens
64. True
65. *A*–Condom, *B*–Copper-T (Cu-T),
    C–Implant, *D*–Vasectomy
66. Process is tubectomy and the structure involved in Fallopian tube.
67. (i) Uterus
68. False. Surgical sterilisation techniques have poor reversibility.
69. Fever is odd one out. Rest of the options are the mild side effects of using contraceptives.
70. Medical Termination of Pregnancy (MTP).
71. Voluntary
72. (i) 40-50 million, (ii) 12 weeks, (iii) 1971, (iv) 2017, (v) Two
73. (ii) MTPs are misused for female foeticide.
74. MTP performed by unqualified quacks, which are unsafe.
75. Sexually Transmitted Diseases (STDs)
76. Venereal Diseases (VDs) or Reproductive Tract Infection (RTI) or Sexually Transmitted Infection (STI).
77. Amoebiasis is odd one out. Rest of the options are STDs.
78. Only (i)
79. Chlamydiasis is odd one out. Rest of the options are non-curable STDs.

80. True

81. 15-24 years age group

82. *A* - Unknown/Multiple

B - Condoms

83. False. Nematodes never cause STDs.

84. Ecotopic pregnancy and abortion.

85. *Treponema pallidum* bacterium

86. True

87. AIDS

88. Enzyme Linked Immuno Sorbent Assay (ELISA)

89. A-3, B-4, C-2, D-1

90. Infertility

91. Assisted Reproductive Technologies (ARTs)

92. False. IVF stands for *In Vitro* Fertilisation.

93. Fallopian tube, Zygote Intra Fallopian Transfer (ZIFT)

94. ZIFT, IUT

95. PID or Pelvic Inflammatory Disease is odd one out. Rest of the options are assisted reproductive technologies.

96. Gamete Intra Fallopian Transfer

97. Intra Cytoplasmic Sperm Injection (ICSI)

98. Vagina

99. A-2, B-1, C-4, D-3

100. True

CHAPTER FIVE

PRINCIPLES OF INHERITANCE AND VARIATION

Topic 1 : Mendel's Laws of Inheritance

1. Select the correct term from the two options provided in the brackets regarding the statements.

   (i) Gregor Mendel conducted ______________________ (linkage/hybridisation) experiments on garden peas for seven years (1856-1863).

   (ii) The characters investigated by Mendel in garden pea were manifested as ______________ (two/four) opposing traits.

2. A true breeding line exhibits stable trait inheritance for several generations.

   True ☐ False ☐

3. Mendel selected four true breeding characters of pea plant for his experiments.

   True ☐ False ☐

4. How many characters given below are dominant in pea plants?

Rounded seed, Terminal flower position, Yellow pod colour, White flowers, Yellow coloured seed

   ____

5. Choose the most appropriate answer.

   (i) Among the following traits of pea plant, which one was not studied by Mendel?

   (a) Seed shape (b) Seed size

   ____

(ii) Inflated pod shape of pea is

(a) Dominant trait (b) Recessive trait

____

6. Which is correct about traits chosen by Mendel for his experiment on pea plant?

Terminal pod was dominant, Constricted pod was dominant, Green coloured pod was dominant, Tall plant was recessive

____

Topic 2 : Inheritance of One Gene

7. Fill in the blanks.

(i) The alternate forms of a gene at same locus on homologous chromosomes are ______________________________

(ii) Cross between genetically different parents yield generation of hybrids called ______________________________

(iii) ______________ are the hybrids which contain contrasting alleles.

8. The Mendel crossed true breeding tall and dwarf plant varieties in his experiment. Tallness was the dominant character and dwarfness was recessive. The recessive character appeared in

____

9. *Rewrite the statements by changing the underlined words.*

(i) While studying inheritance of one gene in pea plant, Mendel found both parental traits appear in F_2-generation in the proportion <u>1 : 1</u>.

____

(ii) The <u>recessive</u> trait always appear (s) in F_1-generation in a monohybrid cross.

____

10. How did Mendel obtained recessive (dwarf) character in F_2-generation?

____

11. In the given diagram, what do the steps I and II represent?

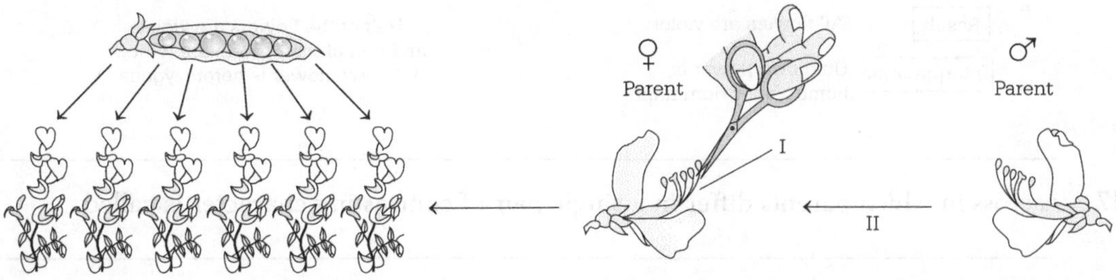

I.______________________ II. ______________________

12. During his experiments, Mendel used the term factor for ____________

13. Give scientific term for the following statements.

(i) A graphical representation to calculate the probability of all possible genotypes of offsprings in a genetic cross. ____________

(ii) Cross between the F_1 hybrid and recessive parent. ____________

(iii) Cross between the F_1 hybrid and any of the parent. ____________

14. Choose between the two options to answer the questions specified in the brackets.

(i) 1 : 2 : 1 or 1 : 1 (What is the genotypic ratio of cross between pure tall and dwarf plants?)

____

(ii) Genotype or Linkage (Which of the following is predicted using test cross?)

____

15. Choose the incorrect match.

(i) Phenotype – Physical appearance of an organism

(ii) Genotype – Expressed genes

(iii) Homozygous – Unidentical alleles of a gene present at the same locus

(iv) Heterozygous – Genes of an allelic pair are not same

____

16. What does the given cross depict?

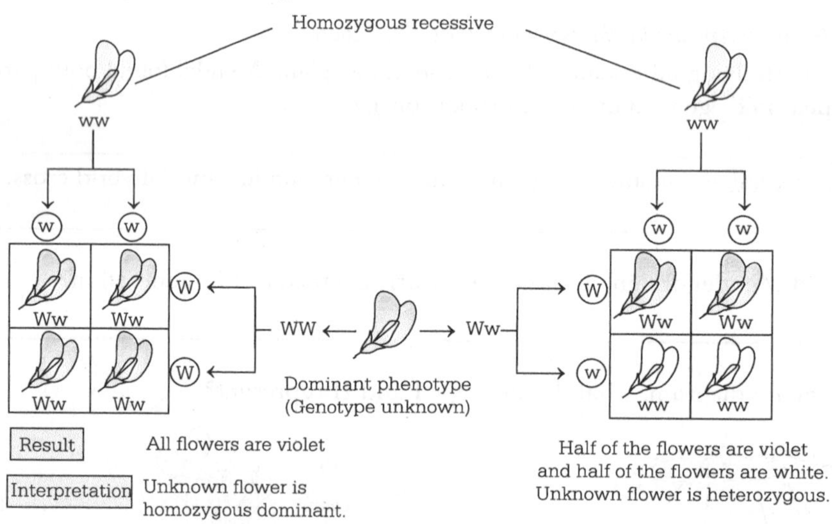

____

17. A cross in which parents differ in a single pair of contrasting character is called

____

18. The phenotypic and genotypic ratio of a monohybrid cross in F_2-generation is

____

19. Law of dominance and law of segregation are based on the inheritance pattern in monohybrid cross.

True ☐ False ☐

20. When alleles of two contrasting characters are present together and one of the character expresses itself during the cross, while the other remains hidden gives the

____

21. The law of dominance is applicable in inheritance of <u>starch grain size</u> in pea. (*Rewrite the statement by changing the underlined word.*)

____

22. Which Mendelian law explains the presence of only one allele in a gamete?

____

23. Mendel's law of segregation is best illustrated by ______________ (Bb × Bb / TT × tt).

24. The law based on fact that the characters do not show any blending and both the characters are recovered as such in F_2-generation although one character was absent in F_1-progeny, is ____

25. Mendel crossed tall and dwarf plants. In F_2-generation, the observed ratio was 3:1 (tall : short). From this result, he deduced

Law of dominance, Law of independent assortment, Law of segregation, Incomplete dominance

____

26. Write the genotype and phenotype of F_1-generation (R–dominant and red, r–recessive and white) in the given cross of snapdragon plant showing incomplete dominance.

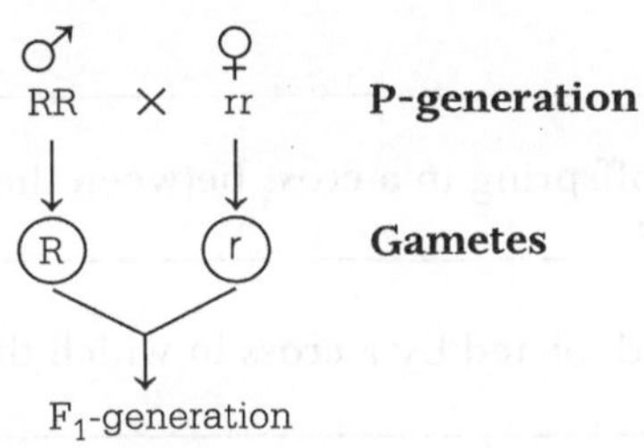

____

27. In ______________________________, one allele function as normal, while another allele may function as normal, non-functional or less efficient.

28. Complete the following flowchart, depicting the inheritance of flower colour in snapdragon.

RR (Red) × rr (White)
↓
F_1-generation Ⓐ (Give phenotype)
↓ Selfing
F_2-generation Phenotypic Ratio-Ⓑ
Genotypic Ratio-Ⓒ

A.______________ B.______________ C. ______________

29. The __________ trait in case of incomplete dominance is seen due to the non-functional enzyme produced by modified gene or absence of an enzyme.

30. The probability of yielding red snapdragon if a pink snapdragon is crossed to a white snapdragon is ______________ (1/4 / zero).

31. Incomplete dominance is similar to codominance in having identical ______________ ratio.

32. TtRr represents heterozygous tall, heterozygous pink plant. If this plant is self-crossed then

T–dominant, t–recessive, R–dominant, r–recessive

Choose the correct option.

(i) 25% plants have red flowers.
(ii) 1% plants have white flowers.
(iii) 25% plants have pink flowers.
(iv) 50% plants are tall.

____

33. What will be the ratio of offspring in a cross between the red coloured and pink coloured flowers of *Mirabilis jalapa*? ______________

34. Which Mendelian idea is depicted by a cross in which the F_1-generation resembles both the parents? ______________

35. Match the following columns.

Column I	Column II
A. Law of dominance	1. F_1 resembles both the parents
B. Incomplete dominance	2. F_1 resembles dominant parent
C. Codominance	3. F_1 exhibits blending inheritance

A.______________ B.______________ C. ______________

36. The ABO blood groups are controlled by c-gene.
(*Rewrite the statement by changing the underlined word.*)

____

37. Out of the three alleles of gene I, the sugar polymers on the plasma membrane of RBCs is controlled by how many alleles? ____________________

38. In human blood group, inheritance __________ are codominant and __________ is recessive allele.

39. *Rewrite the statements by changing the underlined words.*

(i) ABO blood grouping in humans is an example of incomplete dominance.

____

(ii) The genotype of AB blood group is ii. ____________________

40. ABO blood grouping is a good example of multiple alleles.

True ☐ False ☐

41. Complete the table showing genetic basis of blood groups in human population.

Alleles of Parent 1	Alleles of Parent 2	Blood types of offspring
I^A	________	A
I^B	I^A	________
i	i	________

42. Give an example of the production of multiple effects by a single gene in pea plant.

____

43. Heterozygous starch synthesising gene in pea plant produce intermediate sized starch grain due to incomplete dominance.

True ☐ False ☐

Topic 3 : Inheritance of Two Genes

44. How many round green and wrinkled yellow seeds are produced when round yellow seeds are crossed with wrinkled green seeds?

____

45. The number of gametes formed in a typical dihybrid cross is ____________________

46. How many types of gametes can be produced by diploid organism who is heterozygous for four loci? ____________________

47. In a dihybrid cross, Mendel found that dominant and recessive traits of pea plant were segregate in 3 : 1 ratio in F_2 -generation.

Yes ☐ No ☐

48. Which Mendelian law is based on the study of dihybrid cross?

____

49. In a trihybrid cross, Aa Bb Cc × Aa Bb Cc, what would be the

(i) Number of phenotypes ______________________________

(ii) Number of genotypes ______________________________

50. The types of gametes formed by the genotype RrYy are ______________ .

51. In law of independent assortment, how many factors are involved (for a dihybrid cross)?

____

52. In Mendel's experiments with garden pea, round seed shape (RR) was dominant over wrinkled seeds (rr) and yellow colour (YY) was dominant over green colour (yy). What are the expected phenotypes in the F_1-generation of the cross RRYY × rryy?

____

53. In cross, between pure breeding pea plants having yellow round (YYRR) and green wrinkled (yyrr) seeds, find out the total seeds (plants) having yellow colour in F_2-generation.

______________________________ .

54. In a cross, between plants having yellow round (YYRR) and green wrinkled (yyrr) seeds, what will be the ratio between seeds having yellow and green seed colour?

____

55. Match the following columns.

	Column I (Genetic crosses)		Column II (Phenotypic ratios)
A.	Test cross	1.	9 : 3 : 3 : 1
B.	Monohybrid cross	2.	1 : 1
C.	Dihybrid cross	3.	3 : 1

A.______________ B.______________ C. ______________

56. When two pairs of traits are combined in a hybrid, segregation of one pair is independent of the other pair. Is this statement has universal applicability?

Yes ☐ No ☐

57. Mendel published his work on inheritance of characters in ____________ but it remains unrecognised till ______________ .

58. Choose the odd one out with respect to the rediscovery of Mendel's work.

de Vries, Correns, Von Tschermak, Reginald C Punnett

____

59. Which two scientists used chromosome movement to explain Mendel's laws in 1902?

____

60. The two alleles of a gene pair are located on the non-homologous sites on homologous chromosomes.
(Rewrite the statement by changing the underlined word.)

____

61. Read the following statements.
(i) Both chromosomes and genes occur in pairs.
(ii) Pairing and segregation of a pair of chromosomes lead to the segregation of allelic pair they carry.

Which theory in genetics justify these postulates?

____

62. Experimental verification of Sutton and Boveri's theory was given by Thomas Hunt Morgan.

True ☐ False ☐

63. The experimental material used by Morgan for his studies was _______________ (*Drosophila melanogaster / Pisum sativum*).

64. Out of the listed characteristics, how many are found in *Drosophila melanogaster*?

Hermaphrodite, Several hereditary variations, Two weeks life cycle, Easily grown in laboratory

____

65. Choose the incorrect pairing among the following.
(i) Sutton and Boveri – Chromosomal theory
(ii) Walter and Boveri – Behaviour of chromosome during cell divisions
(iii) TH Morgan – Mutation
(iv) Henking – Barr bodies

____

66. Linked genes that were observed by Morgan were present on

____

67. Match the following columns.

	Column I		Column II
A.	Recombination	1.	Non-parental gene combinations
B.	Linkage	2.	Coloured bodies
C.	Chromosomes	3.	Physical association of genes

A.__________ B.__________ C. __________

68. Read the following characteristics.

(i) Violation of the law of segregation.

(ii) High proportion of parental gene combinations.

(iii) Deviation of F_2 ratio from 9 : 3 : 3 : 1 ratio.

Which genetic term uphold all the listed characteristics?

____

69. Linked genes are situated on the ______________________________ (single chromosome/homologous chromosome).

70. Given below is the percentage of recombinant between two genes combinations in *Drosophila*.

Gene combination	% Recombination
A. White eyed and yellow bodied	1.3
B. White eyed and miniature winged	37.2

Which gene combination exhibits high linkage?

____

71. A unit of distance between genes on chromosomes representing 1% cross over in genetic map is ______________________________

72. Genetic maps were first developed by ______________________________

73. The distance between two genes on same chromosome is directly proportional to the strength of linkage between them.

True ☐ False ☐

74. The map distance between various genes is given below.

- *P* and *Q* = 4
- *P* and *R* = 11
- *Q* and *R* = 7

Determine the order of genes on the linkage map. ______________________________

75. Choose the odd one out with respect to linkage experiment.

AaBb × AaBb,	AAbb × AaBB,
AABB × aabb,	aaBB × aaBB

____

76. Match the following columns.

Column I (Scientists)	Column II (Discoveries)
A. Gregor Mendel	1. Chromosomal theory of inheritance
B. Sutton and Boveri	2. Law of inheritance
C. Henking	3. Linkage
D. Morgan	4. Discovered X-body
E. de Vries	5. Mutation

A.________________ B.________________ C. ________________ D. ________________
E. ________________

77. Match the following columns.

Column I	Column II
A. ABO blood groups	1. Dihybrid cross
B. Law of segregation	2. Monohybrid cross
C. Law of independent assortment	3. Base pairs substitution
D. Gene mutation	4. Multiple allelism

A.________________ B.________________ C. ________________ D. ________________

Topic 4 : Polygenic Inheritance and Pleiotropy

78. The polygenic traits are controlled by

One gene, Two genes, Three or more genes, Mutant genes

____

79. In polygenic traits (*Choose the correct one*).

(i) are influenced by environment.

(ii) Effect of each allele is not expressed.

____

80. Human skin colour is a polygenic trait in which all dominant and all recessive alleles, respectively show ________________________ phenotype.

81. In human skin colour inheritance, the genotype with three dominant and three recessive alleles will produce intermediate skin colour.

True ☐ False ☐

82. Read the following characteristics and identify the gene.

(i) It is a single gene which exhibits multiple phenotypic expressions.

(ii) It shows effect on metabolic pathways, so as to produce various phenotypes.

____

83. Phenylketonuria is caused due to

Pleiotropy, Multiple alleles, Codominance, Incomplete dominance

____

84. Phenylketonuria in human (*Choose the correct one*).

(i) do not manifest through phenotypic expressions.

(ii) is characterised by mental retardation, hair reduction and skin pigmentation.

____

85. Which one of the following pairs is wrongly matched?

(i) XO type of sex-determination – Grasshopper

(ii) ABO blood grouping – Codominance

(iii) Starch synthesis in pea – Polygenic trait

(iv) TH Morgan – Linkage

____

86. Match the following columns.

Column I (Terms)		Column II (Descriptions)
A. Dominance	1.	Many genes govern a single character
B. Codominance	2.	In a heterozygous organism, only one allele expresses itself
C. Pleiotropy	3.	In a heterozygous organism, both alleles express themselves fully
D. Polygenic inheritance	4.	Mutation in gene coding phenylalanine hydroxylase

A.__________ B.__________ C. __________ D. __________

Topic 5 : Sex-Determination

87. Henking found X-body in insects while studying ______________ (spermatogenesis/oogenesis).

88. Sex-determination is controlled by __________ and the remaining chromosomes which are not involved in sex-determination are called __________.

89. Choose the incorrect pair with respect to sex-determination in different organisms.

(i) Grasshopper – XO type

(ii) Birds – ZZ-ZW type

(iii) *Drosophila* – XX-XO type

(iv) Human – XX-XY type

____

90. Choose the incorrect pair amongst the following.

(i) Male bird – Homogametic

(ii) Female bird – Heterogametic

(iii) Male *Drosophila* – Homogametic

____

91. Give an example of XO type of sex-determination.

____

92. The number of chromosomes in male and female grasshopper is same.

True ☐ False ☐

93. Among the given male organisms, which one is homogametic?

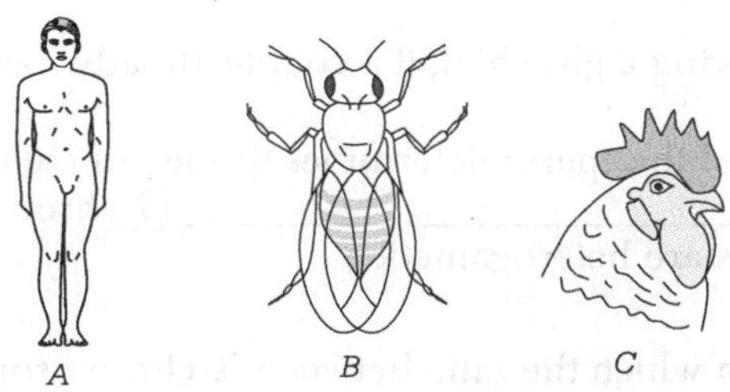

____

94. Female *Drosophila* contains the same sex chromosomes as in female grasshopper and human.

True ☐ False ☐

95. Refer to the genetic cross given below.

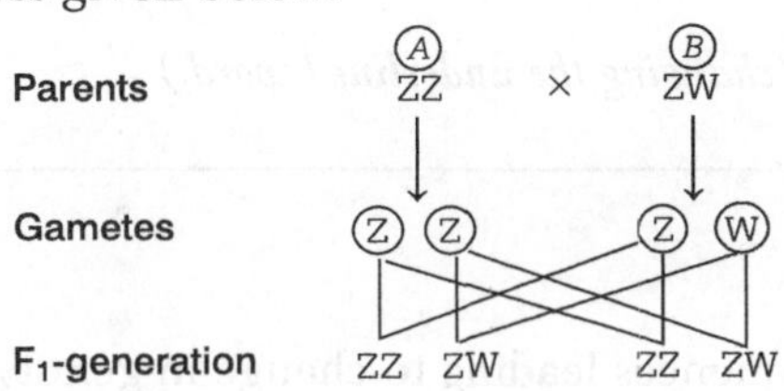

What is the gender of parents, marked as *A* and *B* ?

A.______________ B.______________

96. Match the following columns.

Column I (Organisms)	Column II (Sex-determination mechanisms)
A. Honeybee	1. ZZ-ZW
B. Pigeon	2. Haplodiploidy
C. *Drosophila*	3. XO-XX
D. True bugs	4. XX - XY

A.______________ B.______________ C. ______________ D. ______________

97. Choose the odd one out.

XY, XO, ZW

____

98. A human male contains the karyotype ________________ and a human female has ____________ chromosomes.

99. The number of chromosomes in females and males honeybees are ________________

100. How many pairs of chromosomes are exactly same in human males and females?

____

101. Heterogametes in humans are formed during ____________________________ (spermatogenesis/oogenesis).

102. The probability of having a girl child, if a couple already has two sons is ____________

103. The genetic makeup of the sperm determines the sex of child because ______________________________________ (Y-chromosome is smaller than X-chromosome/males are heterogametic).

104. Name the organism in which the ratio between X-chromosome and autosomes determine the sex.____

105. The unfertilised eggs in honeybees develop into <u>queen</u>. (*Rewrite the statement by changing the underlined word.*)

____

106. In honeybees, male and female gametes are produced through <u>meiosis and mitosis</u>, respectively. (*Rewrite the statement by changing the underlined word.*)

____

Topic 6 : Mutation

107. Alternation of DNA sequences leading to change in genotype and phenotype of organism is called ____________________________________

108. Mutation arising due to the change in a single base pair of DNA is called <u>frame shift</u> mutation. (*Rewrite the statement by changing the underlined word*).

____

109. Sickle-cell anaemia is a classical example of ____________________________

110. Give the collective term for these types of mutations.

Insertion, Duplication, Deletion

____

111. An example of mutagen is ______________________________(UV rays /cancer cells).

Topic 7 : Genetic Disorders

112. Give the scientific term for the analysis of traits in several generations of a family to study the inheritance pattern of a particular trait. ______________________

113. Pedigree analysis employs the use of ______________________ (Punnett square/Standard symbols).

114. How many symbols represent affected individuals among those listed below?

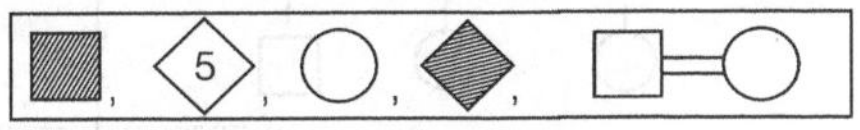

____

115. Which symbol is used to represent the consanguineous mating in pedigree analysis?

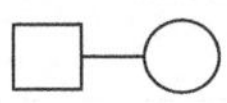 or 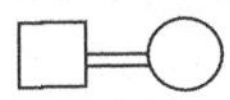

A *B*

____

116. In pedigree analysis, males are represented by a circle and females by square.

True ☐ False ☐

117. Match the following columns.

	Column I (Pedigree symbols)	Column II (Refers to)
A.	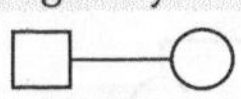	1. Death
B.	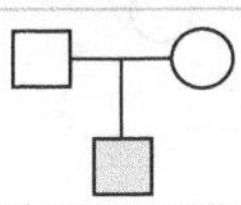	2. Five unaffected offspring
C.		3. Sex unspecified
D.		4. Child affected with disease
E.	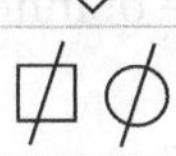	5. Mating

A.__________ B.__________ C. __________ D. __________
E. __________

118. Pedigree analysis is employed to study both Mendelian and chromosomal disorders.

True ☐ False ☐

119. Name the common class of genetic disorders to which all these belong.

Cystic fibrosis, Phenylketonuria, Thalassemia, Colour blindness

____

120. Mendelian disorders are determined by alteration in __________ (single/several) gene(s) and these are transmitted on same lines to offspring.

121. Complete the following analogy.
Autosomal dominant trait : Myotonic dystrophy : : Autosomal recessive trait : __________

122. Given pedigree chart shows the inheritance of polydactyly in humans.

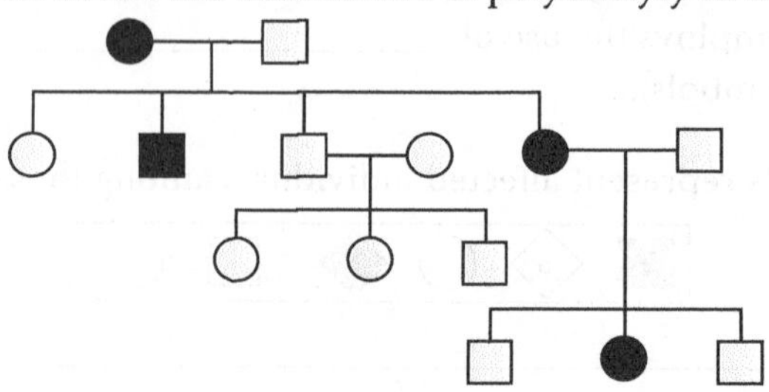

Identify the type of inheritance pattern of this trait.

____

123. How many mutated copies of genes are involved in autosomal dominant and autosomal recessive disorders? ______________________________

124. Given below is the pedigree analysis of fused ear lobes, which is an autosomal recessive trait.

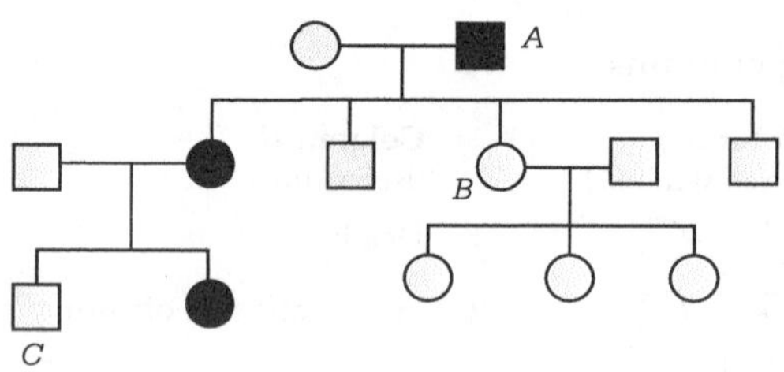

Identify the genotypes of *A*, *B* and *C*.

A.______________ B.______________ C. ______________

125. Refer to the given pedigree analysis and identify the type of inheritance pattern shown in it.

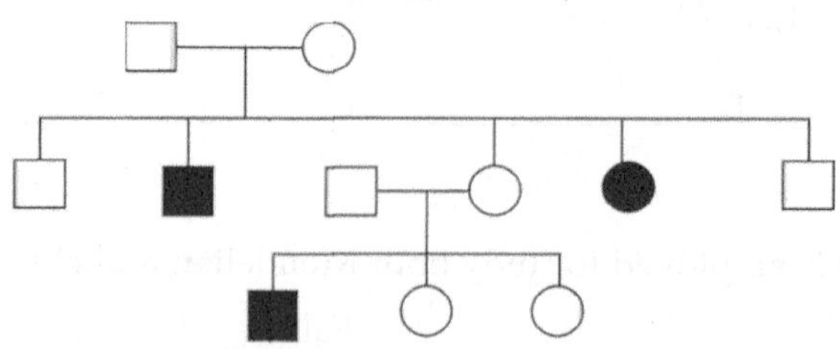

____

126. Choose the odd one out.

Sickle-cell anaemia, Myotonic dystrophy, Colour blindness, Huntington's disease

____

127. Fragile X-syndrome is X-linked dominant disorder.

True ☐ False ☐

128. Colour blindness and haemophilia are sex-linked ____________________ (dominant/recessive) disorders.

129. X-linked recessive traits like colour blindness occur more frequently in males than females because

(i) males receive X-chromosomes from mother.

(ii) females require two-altered copies of genes to get affected.

Most appropriate answer is __________ (Only (i) / Both (i) and (ii))

130. Fathers cannot pass X-linked traits to their sons.

True ☐ False ☐

131. Read the following statements.

(i) Failure to discriminate between red and green colour.

(ii) Mutation in X-linked recessive gene.

(iii) Mostly affect males than females.

These characteristics are indicative of which genetic disorder?

____

132. A marriage between a colourblind man and a normal woman produces ____________________ (all carrier daughters and normal sons/all carrier offsprings).

133. Father of a child is colourblind and mother is a carrier. What is the probability that the child would be colourblind? ____________________

134. Haemophilia is caused due to the deficiency of plasma ______________, which result in non-stop bleeding as blood fails to clot.

135. A female is affected by haemophilia when the mother is at least __________ and father is ______________ .

136. Identify the genotype of person.

(i) Carrier of sickle-cell anaemia. ____________________

(ii) Affected by sickle-cell anaemia. ____________________

137. Refer to the diagram showing the normal β-chain of haemoglobin.

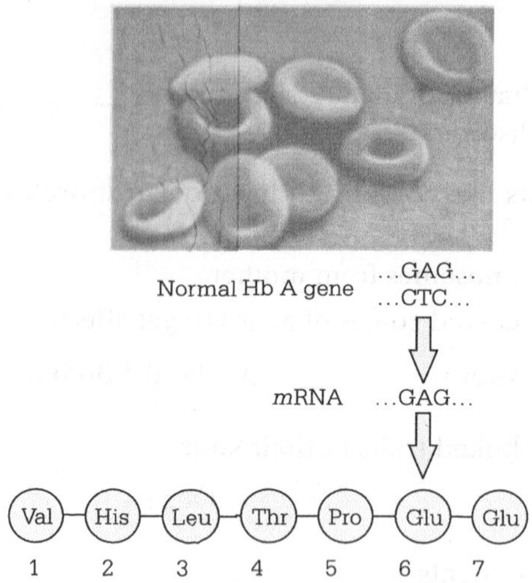

Name the amino acid which gets substituted at position 6 during sickle-cell anaemia.

____

138. Sickle-cell anaemia is caused due to the single base substitution and results in ______________ (GAG → GUG / GUG → GAG).

139. During sickle-cell anaemia, under oxygen tension, the shape of RBCs become <u>spherical</u>. (*Rewrite the statement by changing the underlined word.*)

____

140. An individual affected by phenylketonuria lacks an enzyme that converts the amino acid ______________ into ______________ .

141. Name the acid which gets accumulated in the body of an individual affected by phenylketonuria.______________

142. Mention any two symptoms of phenylketonuria.

(i) ______________ (ii) ______________

143. Match the following columns.

Column I (Genetic disorders)	Column II (Affected chromosomes)
A. Cystic fibrosis	1. 11th chromosome
B. Phenylketonuria	2. 4th chromosome
C. Sickle-cell anaemia	3. 7th chromosome
D. Huntington's disease	4. 12th chromosome

A.______________ B.______________ C. ______________ D. ______________

144. The transmission of thalassemia from parents to offspring requires ________________ (both carrier parents/only one carrier parent).

145. Complete the following flowchart given below (*A-E*).

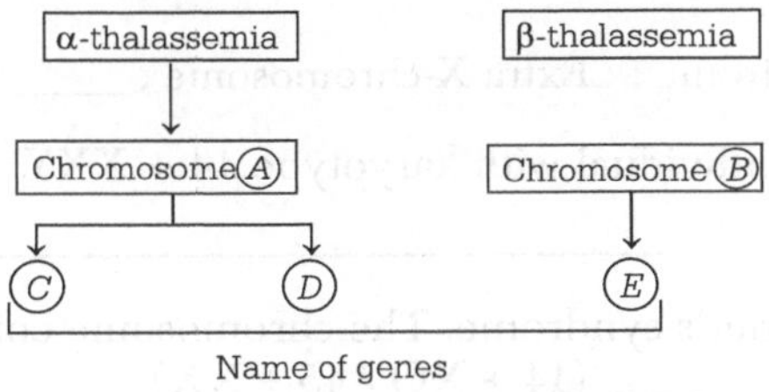

A. ________________

B. ________________

C. ________________

D. ________________

E. ________________

146. Complete the following analogy.

Thalassemia : Either α or β-chain of Hb : : Sickle-cell anaemia : ________________

147. Name the genetic disorders which are caused due to the absence or excess or abnormal arrangement of one or more chromosomes. ________________

148. Complete the following analogy.
Failure of segregation of chromatids : Aneuploidy : : Failure of cytokinesis : ________________

149. Both Down's syndrome and Turner's syndrome are the examples of ________________ (aneuploidy/polyploidy).

150. Choose the odd one out.

Down's syndrome, Turner's syndrome, Klinefelter's syndrome

____

151. Refer to the figure and identify the genetic disorder.

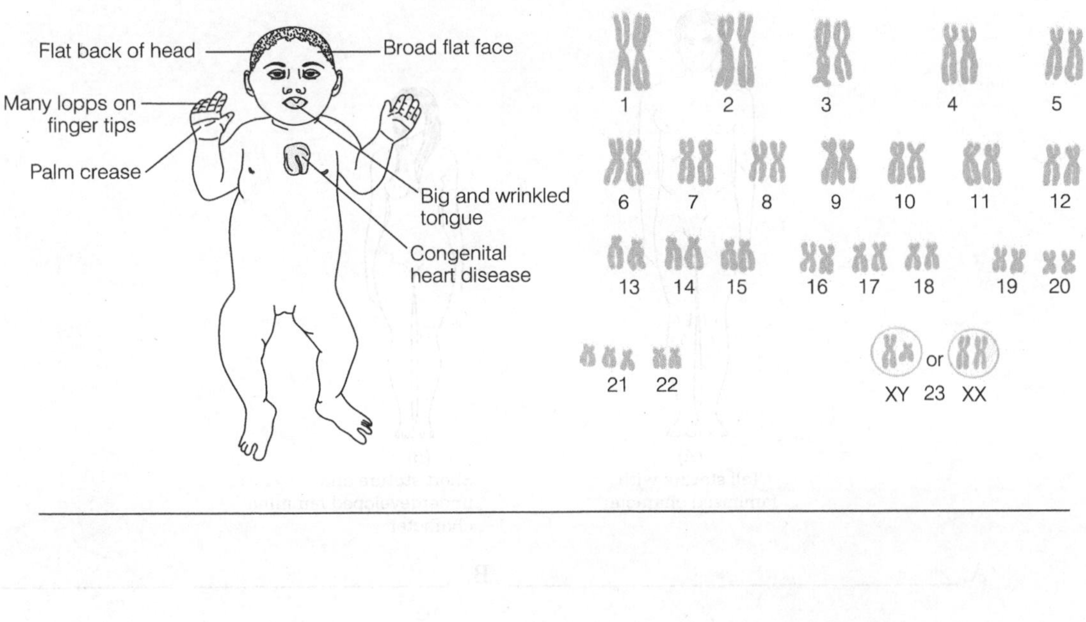

____

152. Give the alternative name of Down's syndrome. ______________________

153. How many chromosomes are found in the genetic disorder described by Langdon Down in 1866? ______________________

154. Complete the following analogy.

Extra chromosome 21 : Down's syndrome : : Extra X-chromosome : ______________

155. Mention any two symptoms in an individual with karyotype 44 + XXY.

____

156. An individual is suffering from Turner's syndrome. The chromosome complement would be ______________________ (44 + XO / 45 + XX).

157. Non-disjunction of 13th chromosome causes Marfan's syndrome.
(*Rewrite the statement by changing the underlined word.*)

____

158. Name the genetic disorder caused due to 18 trisomy. ______________________

159. Match the following columns.

Column I (Genetical traits)	Column II (Name of disorders)
A. Autosomal linked recessive trait	1. Down's syndrome
B. Sex-linked recessive disease	2. Phenylketonuria
C. Inborn metabolic error linked to autosomal recessive trait	3. Haemophilia
D. Additional 21st chromosome	4. Sickle-cell anaemia

A.__________ B.__________ C. __________ D. __________

160. What do the diagrams *A* and *B* represent?

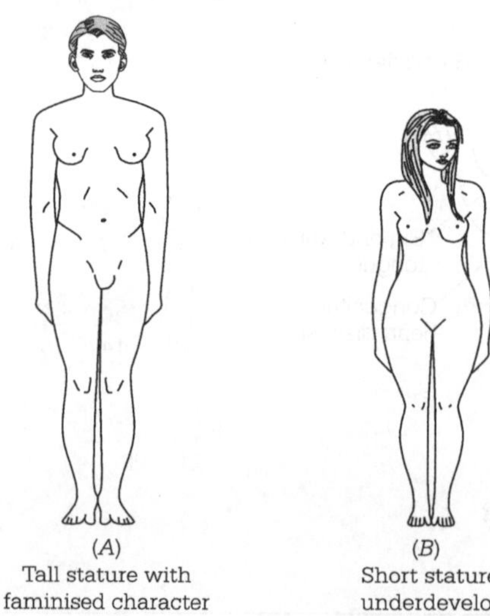

(*A*) Tall stature with faminised character

(*B*) Short stature and underdeveloped faminine character

A.______________________ B. ______________________

Answers

1. (i) Hybridisation
   (ii) Two
2. True
3. False. Mendel selected 14 true breeding characters of pea plant for his experiment.
4. 2 (rounded seed and yellow coloured seed).
5. (i) (b) Seed size
   (ii) (a) Dominant trait
6. Green coloured pod was dominant. Terminal and constricted pods were recessive traits and tall plant was dominant trait.
7. (i) Alleles or Allelomorphs
   (ii) First filial (F_1) generation
   (iii) Heterozygous
8. F_2-generation
9. (i) 3 : 1
   (ii) Dominant
10. By self-pollinating F_1-generation
11. I–Emasculation, II–Pollination
12. Genes
13. (i) Punnett square
    (ii) Test cross
    (iii) Back cross
14. (i) 1 : 2 : 1
    (ii) Genotype
15. (iii) is incorrect because in homozygous condition identical alleles of a gene are present at the same locus.
16. Test cross
17. Monohybrid cross
18. 3 : 1 and 1 : 2 : 1
19. True
20. Law of dominance
21. Seed colour
22. Law of segregation (Second law)
23. Bb × Bb.
24. Law of purity of gametes
25. Law of dominance and segregation
26. Genotype-Rr and Phenotype-pink
27. Incomplete dominance
28. *A* - Pink Rr flowers
    B - 1 : 2 : 1 (Red : Pink : White)
    C - 1 : 2 : 1 (RR : Rr : rr)
29. Recessive
30. Zero. Red colour flowers are homozygous dominant (RR) and they cannot appear in the given cross.
31. Phenotypic and genotypic
32. (i), Selfing of TtRr would give 25% red flower, 25% white flower and 50% pink flower.
33. Red : Pink = 1 : 1
34. Codominance
35. A -2, B -3, C -1.
36. I-gene
37. Two
38. I^A and I^B, i
39. (i) Codominance
    (ii) O
40. True
41. i or I^A, AB, O
42. Starch synthesis in pea seeds.
43. True
44. 3 each, as it is an example of typical dihybrid cross which yield 9 round yellow, 3 round green, 3 wrinkled yellow and 1 wrinkled green seeds.
45. Number of gametes are calculated by the formula 2^n, (where, n = Number of hybrid traits) thus in a dihybrid cross $2^2 = 4$ gametes will be formed.
46. 16 (2^4)
47. Yes
48. Law of independent assortment.
49. Number of phenotypes is 2^n and number of genotypes is 3^n, where n is the number of heterozygous gametes.
    (i) $2^3 = 8$ (ii) $3^3 = 27$
50. RY, Ry, rY, ry
51. 2
52. Only round seeds with yellow cotyledons.
53. 12
54. 3 : 1
55. A-2, B-3, C-1
56. No
57. 1865, 1900
58. Reginald C Punnett is odd one out. He was not involved in the rediscovery of Mendel's results on the inheritance of characters.
59. Walter Sutton and Theodore Boveri.
60. Homologous
61. Chromosomal theory of inheritance.
62. True
63. *Drosophila melanogaster*
64. 3 (*Drosophila* is not a hermaphrodite).
65. (iii) is incorrect because TH Morgan worked as association with linkage.
66. X-chromosome
67. A-1, B-3, C-2.
68. Linkage
69. Single chromosome
70. A-White eyed and yellow bodied.
71. Map unit or centiMorgan.
72. Alfred Sturtevant
73. False. Distance between genes is inversely proportional to the strength of linkage between them.
74. The order of the linkage map will be P-Q-R.

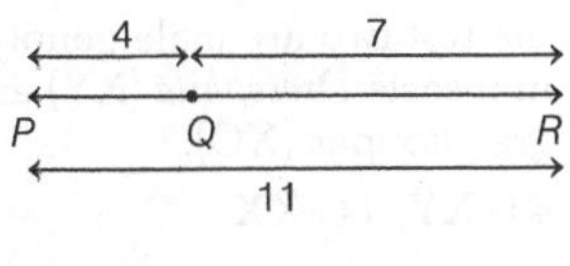

75. AABB × aabb is odd one out as it is the only suitable cross for experiment on linkage as it requires homozygous dominant and homozygous recessive parents. Other crosses are not suitable for this purpose.

76. A-2, B-1, C-4, D-3, E-5

77. A-4, B-2, C-1, D-3

78. Three or more genes

79. (i) is correct. The polygenic traits are influenced by environment in which effect of each allele is additive.

80. Darkest and lightest

81. True

82. Pleiotropic gene

83. Pleiotropy

84. (ii) is correct. Phenylketonuria manifests through phenotypic expressions.

85. (iii) is incorrect match because starch synthesis in pea shows pleiotropism.

86. A-2, B-3, C-4, D-1

87. Spermatogenesis

88. Allosomes, autosomes

89. (iii) is incorrect because in *Drosophila*, XX-XY type of sex-determination is found.

90. (iii) is incorrect because male *Drosophila* is heterogametic.

91. Grasshopper

92. False. Male grasshopper (XO) contains 23 chromosomes, whereas female grasshopper contains 24 chromosomes.

93. *C*. In birds, males are homogametic, whereas females are heterogametic.

94. True

95. A–Male, B–Female.

96. A-2, B-1, C-4, D-3.

97. ZW is odd one out as it is the genotype of female bird whereas the rest two are male genotypes in insects *Drosophila* (XY) and grasshopper (XO).

98. 44+XY, 44+XX

99. 32 and 16, respectively.

100. 22 pairs, i.e. autosomes.

101. Spermatogenesis

102. 50%

103. Males are heterogametic.

104. *Drosophila*

105. Males

106. Mitosis and meiosis

107. Mutation

108. Point

109. Point mutation

110. Frameshift mutations

111. UV rays

112. Pedigree analysis

113. Standard symbols

114. 2 Affected individuals are represented by shaded-shapes.

115. *B* 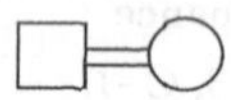

116. False. Male = □, Female = ○.

117. A-5, B-4, C-3, D-2, E-1

118. False. Only Mendelian disorders are traced by pedigree analysis, not chromosomal disorders.

119. Mendelian disorders

120. Single

121. Sickle-cell anaemia.

122. Autosomal dominant trait.

123. One and two, respectively

124. *A*-aa, *B*-Aa, *C*-Aa

125. Autosomal-recessive trait.

126. Colour blindness is odd one out, as it is a sex-linked trait whereas the other three are autosomal linked trait.

127. True

128. Recessive

129. Both (i) and (ii).

130. True

131. Colour blindness.

132. All carrier daughters and normal sons.

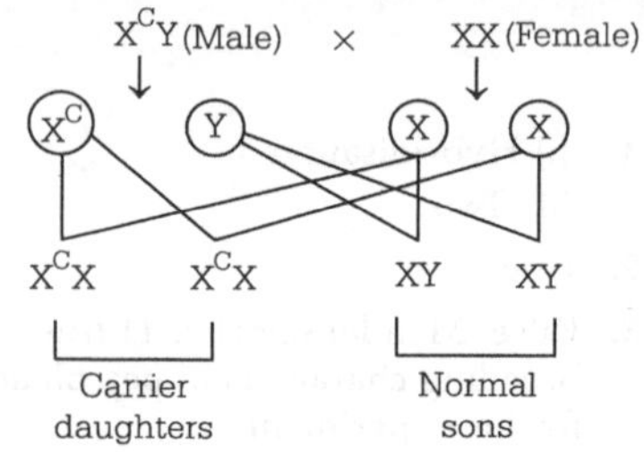

133. 50%. Workout the cross between X^CY (colourblind father) and X^CX (carrier mother).

134. Thromboplastin factors

135. Carrier, haemophilic

136. (i) $Hb^A\ Hb^S$, (ii) $Hb^S\ Hb^S$

137. Valine

138. GAG ⟶ GUG

139. Elongated

140. Phenylalanine, tyrosine

141. Phenylpyruvic acid

142. (i) Hair reduction
(ii) Skin pigmentation.

143. A-3, B-4, C-1, D-2

144. Both carrier parents.

145. *A*–16, *B*–11, *C*–HBA1, *D*–HBA2, *E*–HBB

146. β-chain of Hb

147. Chromosomal disorder

148. Polyploidy

149. Aneuploidy

150. Turner's syndrome is odd one out as it is an example of monosomy whereas the other two represents trisomy.

151. Down's syndrome

152. 21 trisomy

153. 47 chromosomes

154. Klinefelter's syndrome

155. Gynaecomastia, sterile individual

156. 44 +XO

157. Patau's

158. Edward's syndrome

159. A-4, B-3, C-2, D-1

160. *A*–Klinefelter's syndrome
B–Turner's syndrome

CHAPTER SIX

MOLECULAR BASIS OF INHERITANCE

Topic 1 : The DNA : Structure of Polynucleotide Chain

1. Give a scientific term for following sentence.

   The substance that controls the inheritance of traits and also expresses its effects through the formation and functioning of traits. ______________________

2. The nature of the putative genetic material was investigated culminating in the realisation that ______________________ (deoxyribonucleic acids/proteins) are the genetic material, at least in majority of organisms.

3. Name the genetic material in living organisms mentioned below.

   (i) Animals and human beings ______________________ .

   (ii) Viruses ______________________ .

4. DNA mostly functions as a messenger in most living organisms.

   True ☐ False ☐

5. Match the following columns.

	Column I (Organisms)		Column II (DNA contents)
A.	ϕ 174	1.	4.6×10^6 bp
B.	Lambda phage	2.	3.3×10^9 bp
C.	Human DNA	3.	48502 bp
D.	*Escherichia coli*	4.	5386 nucleotides

A. __________ B. __________ C. __________ D. __________

6. DNA is a long polymer of ribonucleotides.
(*Rewrite the statement by changing the underlined word.*)

____

7. In terms of DNA and RNA structure, what is a nucleotide?

____

8. How many of the following are structural components of a polynucleotide chain?

Pentose sugar, Glycerol, Nitrogenous bases, Hexose sugar, A phosphate group

____

9. Identify purines from the following bases.

Adenine, Thymine, Guanine, Cytosine, Uracil, Dihydrouridine

____

10. Thymine is common for both DNA and RNA and cytosine is present only in DNA.
True ☐ False ☐

11. Uracil is present in RNA at the place of _______________ .

12. Two types of pentose sugars are shown below, which of the following is found in RNA and DNA, respectively?

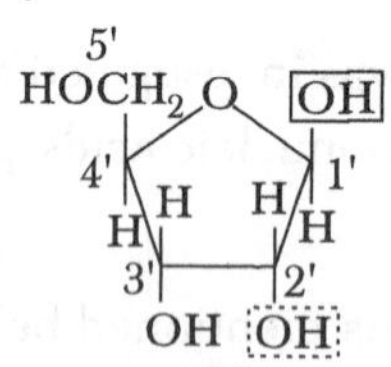

(a)

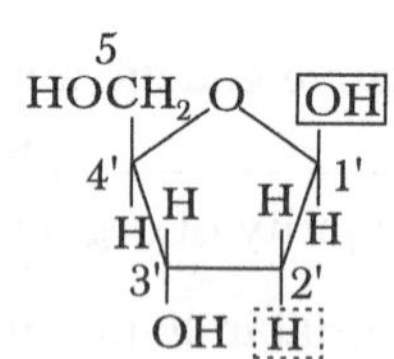

(b)

____

13. With respect to nucleosides which of the following is paired correctly?

Purine–Adenosine, Thymidine/Pyrimidine – Uridine, Cytidine

____

14. Answer the questions based on the figure given below.
(i) The structure given below is _______________ (deoxyadenosine/adenylic acid).

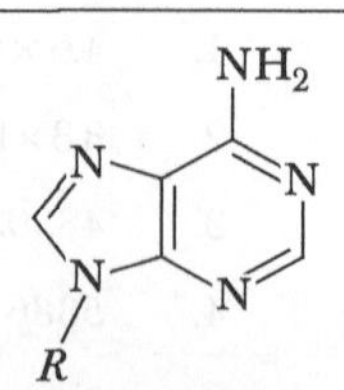

(ii) This structure is formed when nitrogenous base _______________ is linked to the pentose sugar through a _______________ .

(iii) The chemical structure in the figure is a part of a _______________ (DNA/RNA) molecule.

15. A nucleotide is formed when phosphate group is linked to ________________ group of deoxythymidine through ____________________ linkage.

16. In the figure given below, identify the correct labelling of *A-E*.

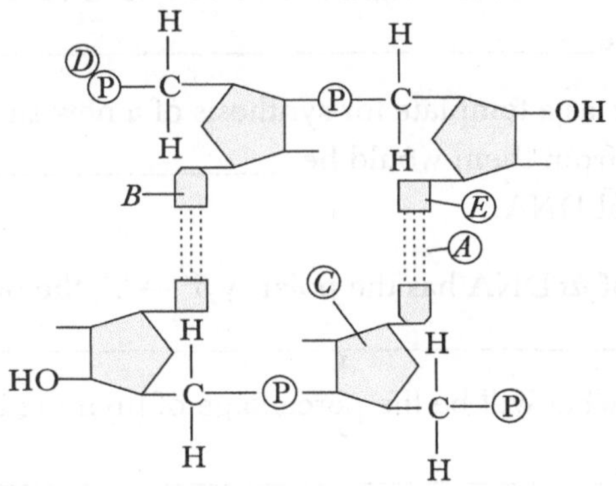

A. ________________

B. ________________

C. ________________

D. ________________

E. ________________

17. State whether the following statements are true or false.

(i) In the polynucleotide chain formed using ribose sugar, a free phosphate moiety is present at 5' end and a free —OH at 3'end.

True ☐ False ☐

(ii) In a dinucleotide, two nucleotides are linked through 3′— 5′ phosphodiester linkage.

True ☐ False ☐

18. The backbone in a polynucleotide chain is formed by ____________________.

19. An additional <u>phosphate group</u> at 2' position in the ribose is found in RNA.
(*Rewrite the statement by changing the underlined word.)*

____

20. In a DNA double helix, the bases are held together by ____________________ (covalent bonds/hydrogen bonds).

21. Match the following columns.

	Column I (Names of scientist)		Column II (Contributions)
A.	Wilkins and Franklin	1.	First identified DNA and named it nuclein
B.	F Meischer	2.	X-ray diffraction of DNA
C.	E Chargaff	3.	Double helix model of DNA
D.	Waston and Crick	4.	Base pairing rules

A. ____________ B. ____________ C. ____________ D. ____________

22. Complete the sentence.

Chargaff's rules of base pairing states that the ratio of ____________________ in all tested organisms.

23. The base pairing confers a very unique property to the polynucleotide chains as it is complementary.

True ☐ False ☐

24. In a *ds* DNA, if the sequence of bases in one strand is 5'- T A C C G A C C A -3', then the sequence of other strand would be ______________________

25. If each strand from a *ds* DNA acts as a template for synthesis of a new strand, the two double-stranded DNA produced from them would be ______________________ (identical/different) to the parental DNA.

26. Which term is used if one chain of *ds* DNA has the polarity $5' \rightarrow 3'$, the other has $3' \rightarrow 5'$ polarity? ______________________

27. If a *ds* DNA has 15% of guanine, what will be the percentage of thymine in it?

____

28. Complete the following analogy with respect to *ds*DNA.

A-T : ______________________ :: G-C : Three H – bonds

29. How is it so that a purine comes opposite to a pyrimidine in *ds* DNA chain?

____

30. Observe the figure given below.

(i) Identify the *A, B, C* and *D.*

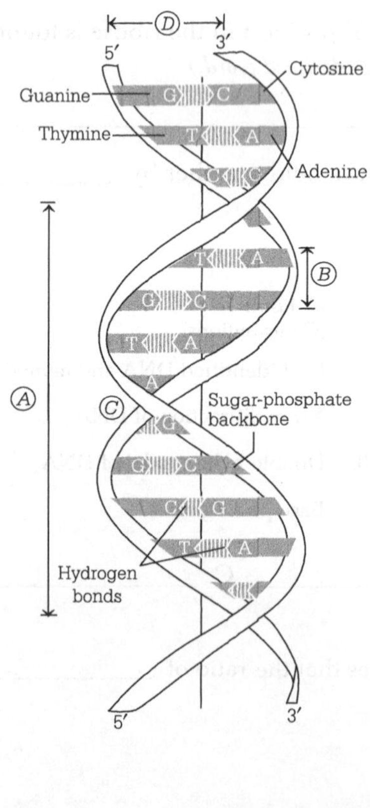

A. ______________________

B. ______________________

C. ______________________

D. ______________________

31. The two chains of the DNA strand are coiled in a ________________ (right/left) handed fashion.

32. The double helix DNA turns after every 10 base pairs at a distance of 3.4 nm. This distance is known as ______________________.

33. The plane of one base pair stacks over the other in double helix, which confers stability of the helical structure.

True ☐ False ☐

34. ______________________ (Watson and Crick/Walther Flemming) were first to suggest that one strand of DNA might act as a template for the synthesis of its complementary strand.

35. Arrange them in the sequence of the flow of genetic information.

(i) Protein, (ii) DNA, (iii) RNA

____

36. In the spaces below, indicate the process that corresponds to each arrow.

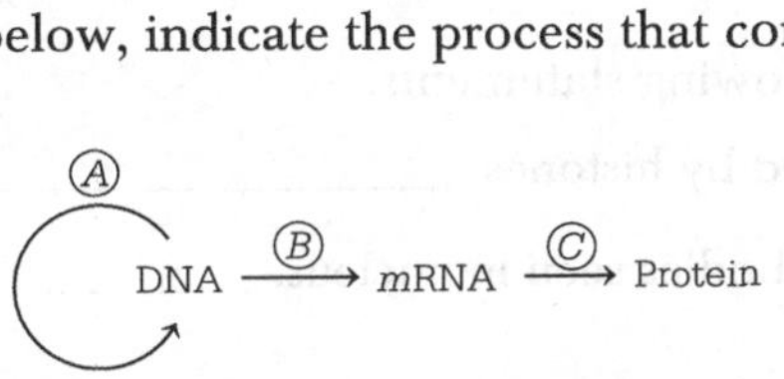

A. ______________________

B. ______________________

C. ______________________

37. Complete the following analogy.
Watson and Crick : Central dogma :: Temin and Baltimore : ______________________

38. The length of DNA double helix in a typical mammalian cell is calculated and it comes out to be approximately 2.2 m. This length is far greater than the dimension of a typical nucleus (approximately 10 superscribe m). So, how is such a long polymer packaged in a cell? (*Answer in one line.*)

____

39. In prokaryotes, the DNA is scattered throughout the cell.

True ☐ False ☐

40. Give scientific term for the poorly defined nuclear region in prokaryotes.

____

41. Nucleosome is made up of how many of the following components?

DNA, Histone core protein, Linker H1, RNA

____

42. Observe the figure given below and answer the following questions.
   (i) Label the parts *A*-*C*.

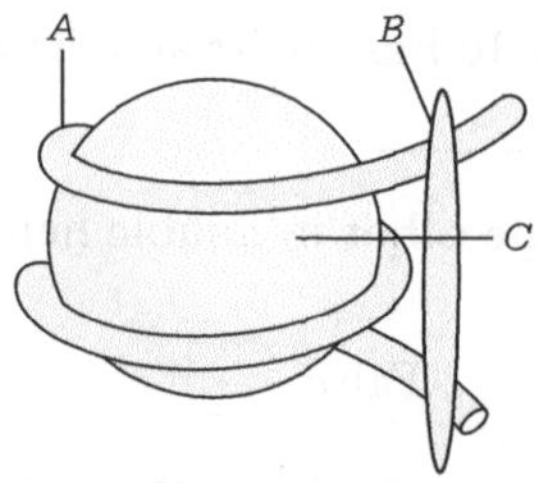

A. ____________
B. ____________
C. ____________

   (ii) The figure represents a ____________ .

43. The nucleosomes in chromatin appear as ____________ (beads on a string/L, V, X shapes) under electron microscope.

44. Basic amino acids help the histones to acquire a ____________ charge on themselves.

45. Lysine and arginine - what is common to both these amino acids? ____________

46. Give a biological term for the following statements.
   (i) A unit of eight molecules formed by histones. ____________
   (ii) Thread-like stained (coloured) bodies seen in nucleus. ____________

47. Complete the following analogy.
   Nucleosome : ____________ :: Histones : Octamer

48. The nucleosomes in chromatin are seen as ____________ structure when viewed under ____________.

49. Given below are structures associated with the packaging of DNA helix, paired with their related features. Complete the pairing by filling in the missing terms.

(A)	Nucleosomes
Non-Histone Chromosomal (NHC) proteins	(B)
Euchromatin	(C)
(D)	More densely packed and stains dark

A. ____________ B. ____________ C. ____________ D. ____________

50. Complete the following analogy.
   Euchromatin : Transcriptionally active :: ____________ : Transcriptionally inactive

Topic 2 : The Search for Genetic Material and RNA World

51. Complete the following analogy.

Frederick Griffith : __________ :: Mendel : Laws of inheritance

52. F Griffith worked on ______________ *(Streptococcus pneumoniae/Haemophilus influenzae)* while trying to develop a vaccine against pneumonia.

53. The S-strain of *Streptococcus pneumoniae*, produces smooth colonies. What does this mean?

54. Given below are the steps of experiment conducted by F Griffith. Answer the questions based on the information provided below.

S-strain → inject into mice → *A*
R-strain → inject into mice → *B*

(i) What happens at *A* and *B* ? ____

(ii) Based on the above results, which strain is virulent? ______________________

55. S-strain (heat-killed) → Injected into mice → Mice lives. In this case, the mice injected with S-strain lives unlike in the original setup. Why so?

____

56. S-strain (heat-killed) + R-strain (live) → inject into mice → *X*. Here, *X* is ____________ (mice die/mice lives).

57. Griffith concluded that R-strain bacteria was transformed by the heat-killed S-strain bacteria.

True ☐ False ☐

58. The transforming factor in Griffith's experiment was identified as ________________ (genetic material/polysaccharide capsule).

59. The genetic material was thought to be a protein, this assumption was corrected by <u>James Watson</u>.

(Rewrite the statement by changing the underlined word.)

____

60. The biochemical nature of 'transforming principle' in Griffith's experiment was determined as ________________ (nucleic acid/glycolipids).

61. Complete the following analogy.

Proteases : Protein digesting enzyme :: ________________ : RNA digesting enzyme

62. Alfred Hershey and Martha Chase (1952) worked with ____________________ (bacteria/bacteriophage) to prove that DNA is the genetic material.

63. In Hershey and Chase's experiment, the radioactive labelled element used were

____

64. The steps of Hershey and Chase's experiment are shown below.
 (i) Identify the steps *A*, *B* and *C*.

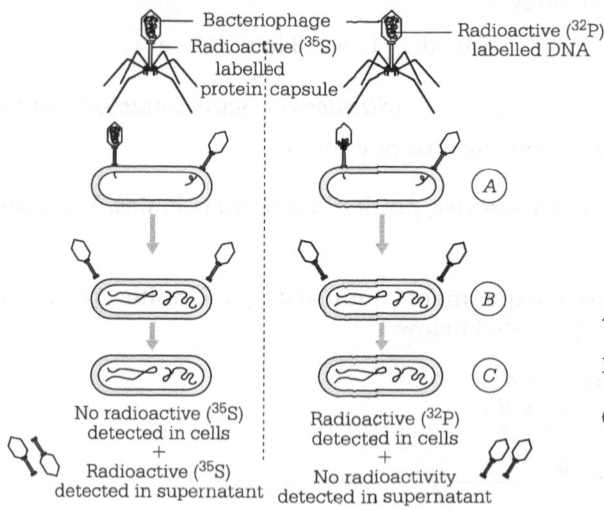

A. ____________
B. ____________
C. ____________

 (ii) In which of the above bacterial cells, will the radioactivity be detected? ____________
 (iii) What conclusion was drawn from this experiment? ____________

65. Match the following columns.

Column I (Scientists)	Column II (Contributions)
A. Frederick Griffith	1. Confirmed DNA as genetic material
B. Hershey and Chase	2. Structure of nucleosome
C. Avery, MacLeod and McCarty	3. Transforming principle
D. Roger Korenberg	4. Biochemical nature of transforming principle

A. ____________ B. ____________ C. ____________ D. ____________

66. Messenger and adapter-these functions are associated with ____________.

67. For a molecule to act as genetic material, it should be capable of ____________ (replication/rapid mutations).

68. <u>Only RNA</u> is capable of directing its duplications.
 (*Rewrite the statement by changing the underlined words.*)

____

69. State whether following statements are True or False.

 (i) Chemically and structurally stable nucleic acid is RNA.
 True ☐ False ☐

 (ii) DNA can express itself in the form of Mendelian characters.
 True ☐ False ☐

70. The first genetic material is ________ (DNA/RNA).

71. A ______________ group present at every nucleotide in RNA is a ______________ group and makes RNA ______________ and easily degradable. RNA is also known to be catalyst, hence reactive.

72. The presence of which nitrogenous base confers additional stability to DNA as compared to RNA?

____

73. Viruses can mutate and evolve faster due to
(i) RNA genome (ii) greater lifespan

____

74. Complete the following analogy.
DNA : Storage of genetic information : : RNA : ______________.

75. Pick the odd one out.

Inheritance, Metabolism, Translation, Splicing

____

Topic 3 : Replication

76. Give a biological term for the following statement.
The type of replication where each DNA molecule would have one parental and one newly synthesised strand. ______________________________

77. Answer the questions associated with the figure given below.

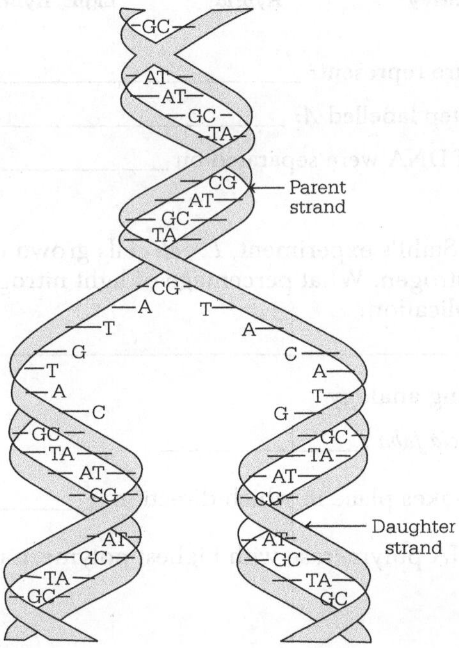

(i) This figure is identified as the semiconservative model of DNA replication.

True ☐ False ☐

(ii) This model of DNA replication was given by <u>Thomas Morgan</u> in 1953.

(*Rewrite the statement by changing the underlined word.*)

____

(iii) In the figure, parent DNA strands undergo _______________ followed by synthesis of a strand. The parent strand acts as a _______________ in this process.

78. DNA replicates semiconservatively was first shown in _______________ *(Streptococcus pneumoniae/ Escherichia coli).*

79. <u>Waston and Crick</u>, experimentally proved semiconservative mode of replication in DNA in 1958.

(*Rewrite the statement by changing the underlined word.*)

____

80. Observe the figure given below and answer the following questions.

Separation of DNA by centrifugation

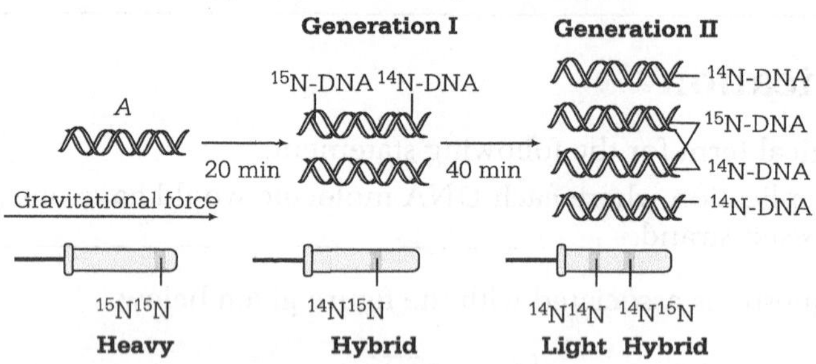

(i) What does the figure represent? _______________

(ii) What happens at step labelled *A*? _______________

(iii) Various samples of DNA were separated on _______________ to measure the densities of DNA.

81. In the Meselson and Stahl's experiment, *E. coli* cells grown on heavy nitrogen were transferred to light nitrogen. What percentage of light nitrogen can be expected after three generations of multiplication?

____

82. Complete the following analogy.

E. coli : $^{15}NH_4Cl$: : *Vicia faba* : _______________

83. Replication of DNA takes place in which direction? _______________

84. Name the type of DNA polymerase with highest polymerisation activity. _______________

85. ______________ are used as substrates in DNA replication, and they also provide energy for polymerisation reaction.

86. What does the efficiency of DNA polymerase activity indicate?

____

87. *E. coli* with ____________ completes the process of replication in ______________.

88. DNA polymerisation rate of DNA polymerase is ____________________ (2000 bp per second/5000 bp per second).

89. 'Polymerases have to catalyse the reaction with high degree of accuracy'. What would happen if there is a mistake in polymerisation? ____________________________

90. In *d*NTPs, the source of energy are the two ________________________ (terminal phosphates/ hydroxyl groups).

91. Match the following columns.

	Column I (Enzymes)		Column II (Functions)
A.	Primase	1.	Addition of nucleotides to growing chain
B.	Helicase	2.	Opening of DNA
C.	DNA polymerase	3.	Unwinding of DNA
D.	Topoisomerases	4.	RNA priming
E.	Gyrase	5.	Removes tension due to supercoiling

A. ______________ B. ______________ C. ______________ D. ______________

E. ______________

92. Give one word answer for statement given below.
A small stretch of DNA helix in which replication occurs. ______________________

93. DNA replication can begin just anywhere on a DNA strand.

True ☐ False ☐

94. During replication, ________ strand is formed because the DNA-dependent DNA polymerases catalyse polymerisation only in one direction, i.e. 5' - 3'.

95. Give the correct heading for the differences given below for templates in DNA replication.

A	*B*
Extends 5' - 3' in the direction of replication fork	Synthesised in 5' - 3' direction but discontinuously
No role of DNA ligase	DNA ligase joins the discontinuous fragments of DNA

A. ______________ B. ______________

96. The discontinuous fragments synthesised on the template with 5 '-3' polarity are called ____________________ (okazaki fragments/ restriction sites).

97. To form a *ds*DNA, it is necessary to join the Okazaki fragments for which single-stranded binding proteins are required.
(*Rewrite the statement by changing the underlined words.*)

____

98. Carefully study the figure given below and answer the questions following.

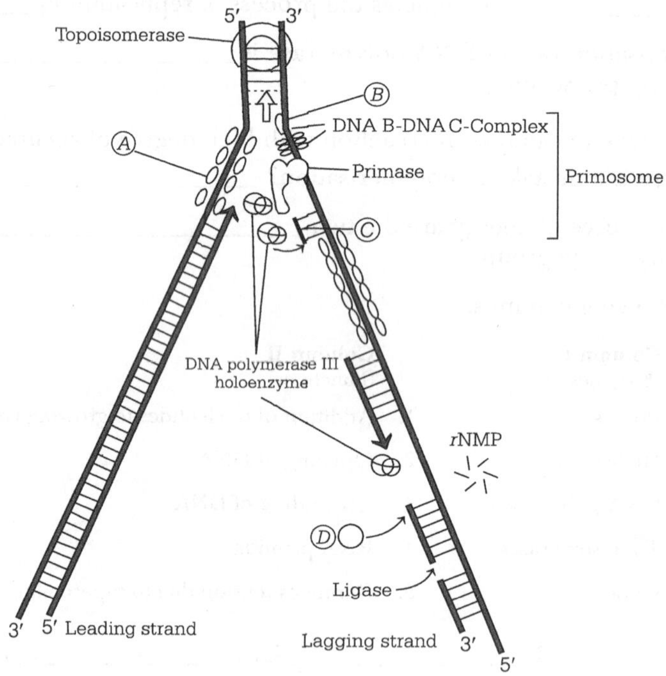

(i) Give a suitable heading for the figure. ______________

(ii) If *A* is not present, DNA single strands can form hairpin structures which retards DNA synthesis. What is *A*? ______________

(iii) '*B*' helps in localised strand separation at *ori.*

True ☐ False ☐

(iv) Why do we add *C* to the replication mixture? ______________

(v) If *D* is not present, replication would not occur because *D* is ______________ .

99. A piece of DNA if needed to be propagated during recombinant DNA procedures, requires a vector because it provides.

____

100. Pick the odd one out.

Helicase, Gyrase, Restriction enzymes, Ligase

____

101. In eukaryotes, the replication of DNA takes place at G_0-phase of the cell cycle.
(*Rewrite the statement by changing the underlined word.*)

____

102. A cell has just doubled up its DNA content but due to a defect in regulatory mechanism they have stopped undergoing division. What is this condition known as?

____

Topic 4 : Transcription

103. Give a scientific term for the following statement.
The process of copying genetic information from one strand of the DNA into RNA.

____

104. Since, RNA is involved, adenine will pair with ______________ while information is being copied from DNA to RNA.

105. In transcription, both segments of DNA are copied into RNA.

True ☐ False ☐

106. If both strands of DNA are transcribed, then two molecules of RNA with similar sequences are produced.
(*Rewrite the statement by changing the underlined word.*)

____

107. Another complication arising from transcription of both DNA strands is the formation of *ds*RNA. This renders the exercise of transcription futile, Why?

____

108. Similar to replication, transcription also occurs bidirectionally.

True ☐ False ☐

109. How many of the following are components of a transcription unit?

Promoter, Replication fork, *Ori*, Structural gene, Terminator, RNA coding gene, Ribozymes

____

110. Complete the following analogy with respect to DNA dependent RNA polymerase.
$3' - 5'$ strand : Template:: $5' - 3'$ strand : ______________________________

111. Template strand → 3' - ATT-CCC-GGC-GCA-TGC-ATG-CAT-GCA- 5'
Coding strand → 5'- TAA-GGG-CCG-CGT-ACG-TAC-GTA-CGT -3'.
Based on this information, predict the sequence of RNA transcribed from above DNA.

____

112. Identify the structures *A, B, C* and *D* in the representation of a transcription unit.

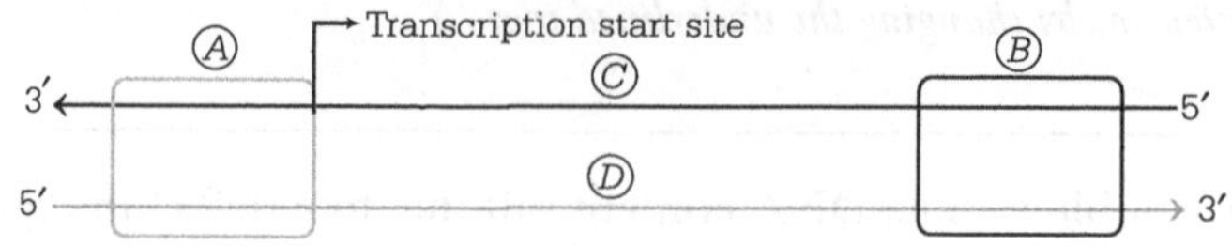

A. ______________ B. ______________ C. ______________ D. ______________

113. With reference to structural genes, indicate the location of

(i) Promoter ______________

(ii) Terminator ______________

114. ______________ is a DNA sequence that provides a binding site for RNA polymerase.

115. ______________ usually defines the end of the process of transcription.

116. A stretch of six nucleotides with consensus sequence 5'- TATAAT-3' in promoter gene is ______________ (RNA coding sequence box/ Pribnow box).

117. Give scientific term for the functional unit of inheritance. ______________

118. Give scientific term for a segment of DNA coding for a polypeptide. ______________

119. Complete the following analogy.

Monocistronic unit : ______________ : : Polycistronic unit : Bacteria

120. The genes in eukaryotes are split. What does this mean?

____

121. Complete the following analogy.

______________ : Coding sequences :: Introns : Non-coding sequences

122. ______________ (Exons/Introns) are said to be those sequences that appear in mature or processed RNA.

123. Match the following columns.

	Column I (Types of RNA)		Column II (Functions)
A.	*t*RNA	1.	Encodes amino acid sequence of a polypeptide
B.	*hn*RNA	2.	Primary transcript from DNA
C.	*m*RNA	3.	Translate the *m*RNA with ribosomal proteins
D.	*r*RNA	4.	Brings amino acids to ribosomes during translation

A. ______________ B. ______________ C. ______________ D. ______________

124. Name the RNA that plays a structural and catalytic role during translation.

____

125. DNA dependent DNA polymerase catalyses transcription of all types of RNA in bacteria. (*Rewrite the statement by changing the underlined word.*)

____

126. To begin the transcription, RNA polymerase recognises nucleotide sequence in ______________ (promoter region / G -C rich region) of the DNA.

127. In prokaryotes, ______________ (single/three) RNA polymerase (s) present.

128. RNA polymerase requires primer and has the ability to repair mistakes in RNA.

True ☐ False ☐

129. How many of the following are components of prokaryotic RNA polymerase?

Core enzyme, Holoenzyme, Enhancers, Coding gene

____

130. The three steps of transcription are depicted below in random order. Rearrange them in their proper sequence.

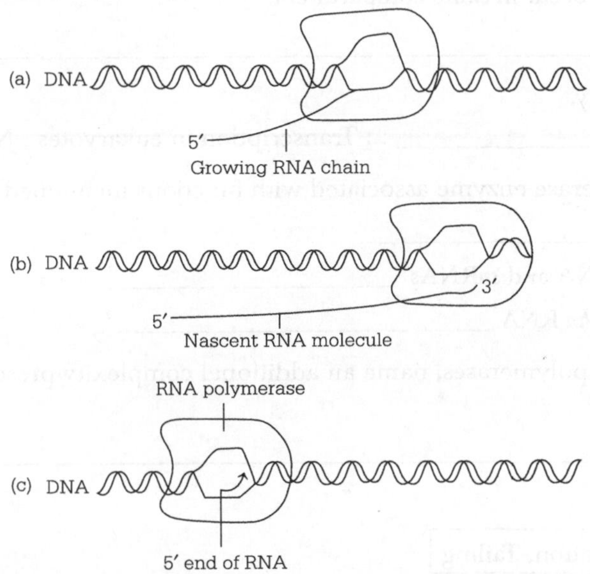

____

131. During initiation, the steps involved are

Promoter recognition

↓

(*A*)

↓

(*B*)

Formation of bonds between the first few ribonucleotides in the nascent RNA chain. Identify *A* and *B*.

A. ________________ B. ________________

132. After the release of ____________ RNA polymerase core enzyme elongates the chain.

133. RNA polymerase uses ________________ (deoxyribonucleoside diphosphates/ribonucleotide triphosphates) to elongate the chain.

134. Complete the following analogy.
σ : Initiation factor :: ________________ : Termination factor

135. In bacteria, translation begin much before the *m*RNA is fully transcribed because

(i) *m*RNA requires processing to be come active.

(ii) transcription and translation occur in same compartment.

____

136. Complete the following analogy.
Transcription in bacteria : ________________ : : Transcription in eukaryotes : Nucleus

137. Name the type of RNA polymerase enzyme associated with functions mentioned below.

(i) Transcribes *r*RNAs ________________

(ii) Transcription of *t*RNA, 5*sr*RNA and *sn*RNAs ________________

(iii) Precursor of *m*RNA and the *hn* RNA ________________

138. Apart from having three RNA polymerases, name an additional complexity present in eukaryotic transcription.

____

139. Pick the odd one out.

Splicing, Capping, Transformation, Tailing

____

140. In post-transcriptional processing, exons are excised and hydrolysed before *m*RNA moves out of the nucleus.

True ☐ False ☐

141. Most eukaryotic pre *m*RNAs are matured by undergoing modifications to their ends? Name these modifications.________________________________

142. Observe the figure given below and answer the questions following.

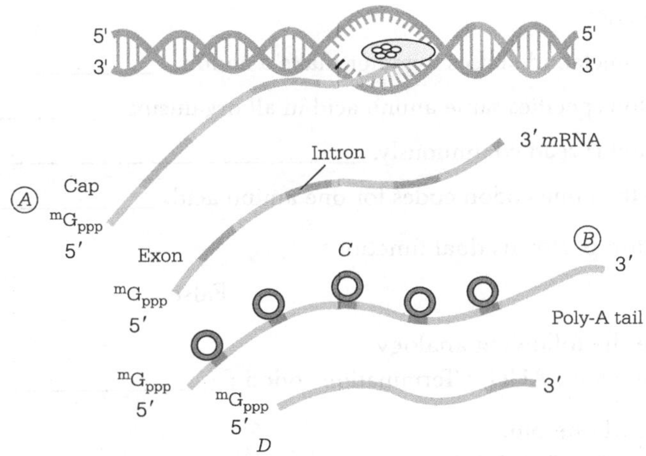

(i) Name the process depicted by the figure. ______________

(ii) The exons and the introns together make the ______________ which is non-functional.

(iii) The introns are removed and exons are joined in a defined order by a process called ______________.

(iv) *hn*RNA undergoes two additional processes for maturation. In ______________, an unusual nucleotide is added to the 5'-end of *hn*RNA. In tailing, ______________ residues are added at 3'-end in a template independent manner by the process called ______________

143. The fully processed *hn*RNA is called exon that is transported out of the nucleus for translation.

True ☐ False ☐

144. Mature eukaryotic *m*RNAs have 5'cap that is a residue of ______________ (2-methylguanosine/7-methylguanosine).

Topic 5 : Genetic Code

145. Give the scientific term for the sequence of nitrogenous bases in *m*RNA coding for single amino acid.

____

146. Complete the table given below.

Scientists	Contributions
George Gamow	(A)
(B)	Synthesised RNA molecules with defined combinations of bases
Marshall Nirenberg	(C)
(D)	Enzymatic synthesis of RNA

A. ______________ B. ______________ C. ______________ D. ______________

147. Mention the characteristic feature of genetic codes associated with the following characteristics.

(i) Three nucleotide bases together make a codon. ______

(ii) A codon specifies same amino acid in all organisms. ______

(iii) A codon is read continuously. ______

(iv) More than one codon codes for one amino acid. ______

148. AUG codon performs dual functions.

True ☐ False ☐

149. Complete the following analogy.
Initiation codon : AUG :: Termination codon : ______ .

150. Pick the odd one out.

GUG, UAG, UGA, UAA

______

151. Degeneracy is due to ______ (first /last) base in the codon.

152. Stop codons are also called ______ or ______ codons.

153. If following is the sequence of nucleotides in *m*RNA, predict the sequence of amino acid coded by it– AUG CUU UUU AAA UGU GGU UUC.

______

154. We know that AUG is an initiation codon, while UAA/UAG/UGA are termination codons. A polypeptide cannot be synthesised if it encounters any of these latter codons. On the basis of this information identify the *m*RNA chain that can be translated completely, ______
(AUG UAU UUC UGC CUC UAG / AUG UUC UCC UGA UGG UUA.)

155. Pick the odd one out.

CUU, CUC, CCA, CUG

______

156. Codon GGU codes for amino acid glycine which is also a function of <u>CGU, UAU and AGG</u>.
(*Rewrite the statement by changing the underlined word.*)

______

157. Give a scientific term for an alternation in the nucleotide sequence of the genome of an organism. ______

158. The change in base sequence alters the phenotype.

True ☐ False ☐

159. Mutations which do not cause any functional change in protein are ______ (non-sense/silent) mutations.

160. A single nucleotide pair is inserted near the 5' end of a protein coding DNA sequence. Its most likely effect will be the production of a protein with
(i) single altered amino acid.
(ii) completely altered sequence.

____

161. Complete the following analogy.
Point mutation : Change in single base :: Frameshift mutations: ____________

162. A classical example of ________ mutation is sickle-cell anaemia.

163. A change of single base pair in the gene for ____________ chain results in the change of amino acid residue from glutamate to ____________, resulting in sickle-cell anaemia.

164. Write the correct bases at the 6th position of DNA, RNA and amino acids of the beta chain of haemoglobin which results in sickle-cell anaemia.

____

165. What is a common consequence of insertion and deletion of base pairs in codons?

____

166. An adapter molecule is *r*RNA.
True ☐ False ☐

167. Who postulated the presence of an adapter molecule for reading the code and to link it to the amino acids?

____

168. In the given figure of *t*RNA,
(i) Identify the labels *A-E*

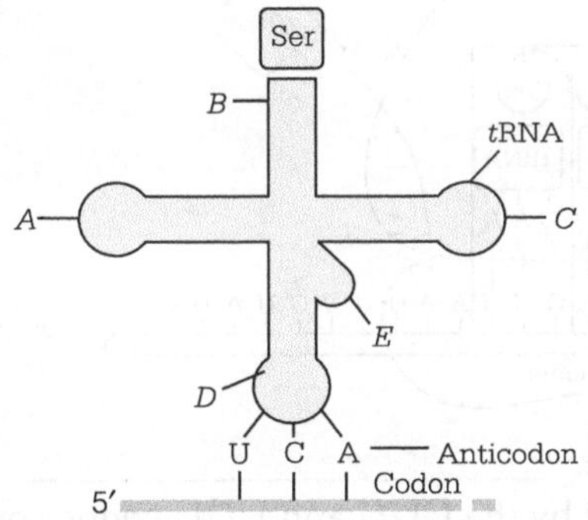

A. ____________
B. ____________
C. ____________
D. ____________
E. ____________

(ii) Write the *m*RNA codon complementary to anticodon on *t*RNA. ____________

169. Complete the sentence and the figure given below

(i) There was a need for an adapter molecule because ________________.

(ii) Identify the parts labelled as *A, B* and *C* in the given figure of *t*RNA (an adapter molecule).

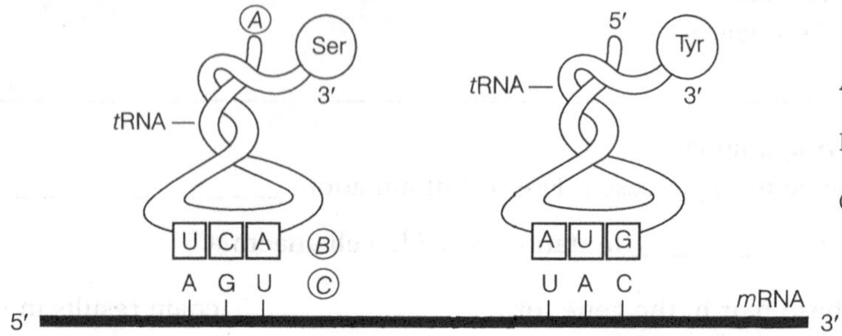

A. ________________

B. ________________

C. ________________

170. The clover leaf model of *t*RNA was given by ________________.

Topic 6 : Translation

171. Give a scientific term for the process of polymerisation of amino acids to form a polypeptide.

____

172. The order and sequence of amino acids are defined by the sequence of bases in the *t*RNA.

True ☐ False ☐

173. During translation, proteins are synthesised by ribosomes using the information on ________________.

174. Observe the figure given below and answer the questions related to it.

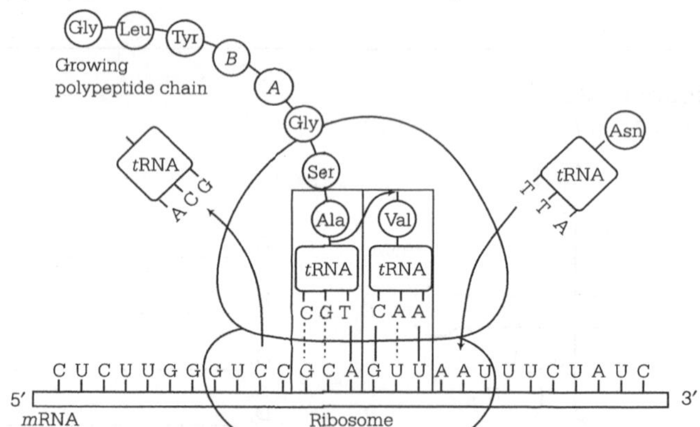

(i) Identify the process shown in the figure. ________________

(ii) Identify the amino acids *A* and *B* coded by the CUU and UCU codons on *m*RNA chain. ________________

175. Given below are the steps involved in the translation process.

Chain formation, Activation of amino acids, Modifications, Initiation, Termination. Rearrange these steps in the correct sequence of protein synthesis.

____

176. Complete the following analogy with respect to ribosomes.

A–site: Accepts charged *t*RNA molecules :: P–site: ________________.

177. How many of the following are essential requirements for the translation process?

Initiation factors, RNA polymerase, Polyribosomes, *hn*RNA, Elongation factors, Translocase, GTPase, *m*RNA, *t*RNA, Aminoacyl *t*RNA synthetase, Amino acids

____

178. Amino acids are activated in the presence of ATP and linked to their cognate *t*RNA-a process commonly called as *X*. Identify *X*.

____

179. Expand UTR. ________________________________

180. The untranslated regions on the *m*RNA have additional sequences which are not translated, thus making this region futile.

True ☐ False ☐

181. 23S *r*RNA in bacteria is the enzyme called <u>ribonuclease</u>.
(*Rewrite the statement by changing the underlined word.*)

____

182. Ribosomes in their inactive state in eukaryotes have two subunits, i.e. a smaller ________________________ and a larger ____________________ units.

183. Given below is the flowchart depicting the steps of translation.

Initiation begins when the ribosomes bind to *m*RNA at the start codon AUG recognised by Ⓐ.

↓

Ⓑ

↓

Ribosomes move from codon-codon along the *m*RNA

↓

Ⓒ

↓

Ⓓ

↓

Polypeptide is released from the ribosome.

Identify *A*, *B*, *C* and *D*.

A. __________ B. __________ C. __________ D. __________

Topic 7 : Regulation of Gene Expression

184. In eukaryotes, gene regulation can be exerted at many levels. Identify these levels and write their names.

____

185. Genes that are continually being expressed in most cells are <u>repressible</u> genes.
(*Rewrite the statement by changing the underlined word.*)

____

186. Refer to the reaction given below in *E. coli*,

Lactose $\xrightarrow{(A)}$ Galactose and Glucose. Identify *A*.

___

187. If the bacteria do not have lactose around them to be utilised for energy source

(i) they would require the enzyme needed for lactose metabolism.

(ii) they would not require enzyme.

___

188. The metabolic, physiological or environmental conditions can regulate the expression of genes.

True ☐ False ☐

189. In a transcription unit, the activity of ______ polymerase at a given promoter is in turn regulated by interaction with ______ proteins which affects its ability to recognise start sites.

190. Give a scientific term for a functional unit of transcription and genetic regulation.

___

191. In an inducible operon, a free repressor cannot bind to the operator.

True ☐ Flase ☐

192. *Lac* operon is a monocistronic structural gene.
(*Rewrite the statement by changing the underlined word.*)

___

193. Answer the questions related to the diagram.

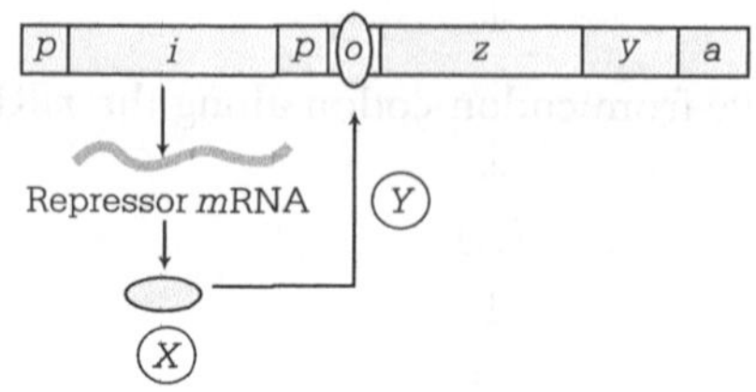

(i) *Lac* operon as illustrated in the figure switched ______ (on/off).

(ii) *X* is the inducer of the given operon. ______

(iii) To which component of operon, *X* is binding? ______

(iv) The elucidation of *lac* operon was carried out by ______ and ______.

194. Observe the figure given below and answer the related questions.

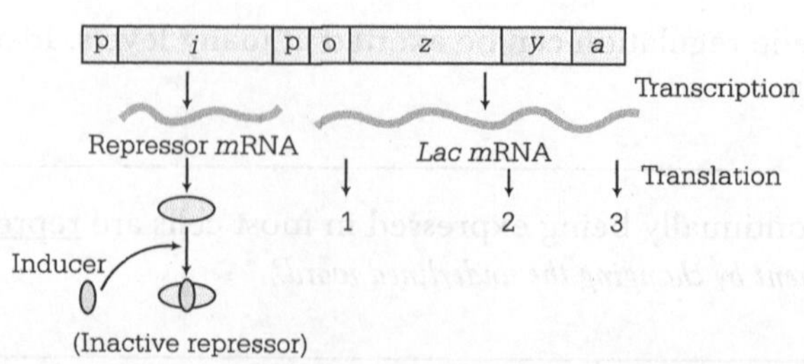

(i) The *lac* operon in figure indicates that lactose is absent in the medium.

True ☐ False ☐

(ii) Name the recognition centre for RNA polymerase in this operon. ________________

(iii) Gene that gives accessibility to RNA polymerase facilitating expression of structural genes is ________________.

(iv) Which gene component produces a repressor protein that binds to the *o*-gene? ________

195. Match the following columns.

Column I (Enzymes)	Column II (Functions)
A. Beta galactosidase	1. Attaches a particular chemical group to target molecules
B. Permease	2. Hydrolysis of lactose sugar
C. Transacetylase	3. Increases permeability of the cell to β-galactosidase

A. ________________ B. ________________ C. ________________

196. Given below are steps involved in *lac* operon.

(A) synthesise repsessor

↓

Lactose induces the operon

↓

Repressor is inactivated by interaction with inducer

↓

(B)

A. ________________

B. ________________

197. Glucose and galactose cannot act as inducers for *lac* operon because they cannot bind repressor.

True ☐ False ☐

____

Topic 8 : Human Genome Project

198. Regulation of *lac* operon by repressor is <u>positive</u> regulation.
(Rewrite the statement by changing the underlined word.)

____

199. The genetic makeup of an organism or an individual lies in his/her DNA sequences.

True ☐ False ☐

200. Name any two techniques of genetic engineering that helped in the launching of human genome sequencing project?

____

201. Expand HGP. ____

202. HGP was launched in year ______________________ (1990/1992).

203. Human Genome Project (HGP) was called a mega project. (*Give reason*)

____

204. Human genome contains approximately __________ base pairs.

205. HGP was closely associated with the rapid development of a new area in biology called biotechnology.

True ☐ False ☐

206. A goal of HGP was to identify all the genes in the human genome which numbered approximately <u>50,000-55,000 genes</u>.
(*Rewrite the statement by changing the underlined word.*)

____

207. The information generated from HGP was stored in monographs.

True ☐ False ☐

208. Human genome project was a ________ (13/17) years project.

209. Name the agencies which coordinated the human genome project.

____

210. What is total estimated cost of human genome project?

____

211. Which of the following was not a goal of HGP?

Improve manpower for data analysis, Store information in databases.

____

212. How many of the following are the organisms whose genome have been sequenced?

Bacteria, Yeast, *Caenorhabditis elegans*, *Drosophila*, rice, *Arabidopsis*, Chimpanzee, Sparrow

____

213. Name the major approaches used in HGP for identifying genes.

____

214. Sequencing the entire genome containing all the coding and non-coding sequences and assigning them to different regions in the sequence of functions is associated with ______________________ (expressed sequence tags/sequence annotation).

215. Steps of sequencing genome are

Total DNA from a cell is isolated and converted into random fragments of relatively smaller sizes

↓

Ⓐ

↓

Amplification of each piece of DNA fragment using Ⓑ

↓

Ⓒ

↓

Ⓓ

These sequences were subsequently annotated and were assigned to each chromosome.

The missing steps are written below, label each of these steps as *A, B, C* and *D*.

Cloning in a suitable host using specialised vectors, Alignment of these sequences, Fragments were sequenced using automated DNA sequencers, PCR

A. ________________ B. ________________ C. ________________ D. ________________

216. Sanger developed method for determination of amino acid sequences in proteins called ________________ .

217. The commonly used hosts for cloning were *Drosophila* and pea for which YAC and BAC vectors are used.
(*Rewrite the statement by changing the underlined words.*)

____

218. The last chromosome to be sequenced in HGP was ________________ .

219. Genetic and physical maps of genome were generated using the information
(i) derived from all parts of the body.
(ii) on polymorphism of restriction of recognition sites and satellite DNA.

____

220. Largest known human gene is ________________ that has been sequenced with 2.4 million bases.

221. What per cent of the human genome codes for proteins?

____

Topic 9 : DNA Fingerprinting

222. Almost 99.9 per cent nucleotide bases differ among all peoples responsible for phenotypic variations.

True ☐ False ☐

223. ___________ sequences in our DNA are used for studying chromosome structure evolution, etc.

224. Complete the following analogy.

Chromosome 1 : 2968 genes :: _______________ : 231 genes

225. Expand SNPs.___________

226. Name a quick way to compare the DNA sequences of any two individuals.

____

227. What forms the basis of DNA fingerprinting technique?

____

228. Do satellite DNA code for any proteins?

____

229. DNA fingerprinting is the basis of paternity testing in case of disputes because

(i) polymorphism in DNA is inheritable.
(ii) SAT DNA is absent in human genome.

____

230. Give a scientific term for variations at genetic level.

____

231. Polymorphisms arise due to mutations.

True ☐ False ☐

232. A mutation can only spread to offspring when it has occurred in <u>somatic cells</u> of sexually reproducing parents.
(*Rewrite the statement by changing the underlined word.*)

____

233. An inheritable mutation occurring at high frequency is called ___________

234. Name the scientist who developed the technique of DNA fingerprinting.

____

235. Expand VNTR. ___________

236. VNTRs are ___________ (minisatellites/microsatellites).

237. Steps of DNA fingerprinting technique are indicated in the flowchart below.

(i) Separation of DNA fragments by electrophoresis.
(ii) Digestion of DNA.
(iii) Transferring of separated DNA fragments to synthetic membranes.
(iv) Isolation of DNA.
(v) Hybridisation using labelled VNTR probe.
(vi) Detection of hybridised DNA fragments by autoradiography.

Arrange these steps in a sequential manner.

____

238. Complete the following analogy with respect to fingerprinting techniques.

Digestion of DNA : ____________________ :: Separation of DNA fragments : Electrophoresis

239. The size of VNTRs varies from 10-20 kb.

True ☐ False ☐

240. Autoradiogram gives many bands of differing sizes. These bands give a characteristic pattern for an individual DNA. It differs from individual to individual in a population except in ____________ .

241. ____________________ (PCR/Autoradiography) has increased the sensitivity of fingerprinting technique.

242. Given below is the electrophoretic reading carried out on a DNA sample found at a site of murder.

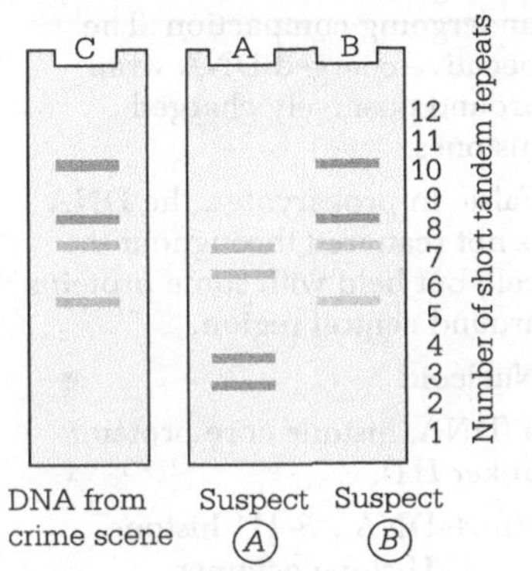

Among the two suspects *A* and *B* which is most likely to have committed the crime.

____

Answers

1. Genetic material
2. Deoxyribonucleic acids
3. (i) DNA, (ii) RNA
4. False. RNA mostly functions as a messenger in most living organisms.
5. A-4, B-3, C-2, D-1
6. Deoxyribonucleotides
7. A nucleotide is a sugar molecule bounded to phosphate group(s) and a heterocyclic base or pentose sugar.
8. 3 (pentose sugar, nitrogenous bases and a phosphate group).
9. Adenine and guanine.
10. False. Cytosine is common for both DNA and RNA and uracil is present only in RNA.
11. Thymine
12. (a) is ribose, thus it is present in RNA, (b) is deoxyribose sugar present in DNA.
13. The correct pair among these is pyrimidine-uridine, cytidine.
14. (i) Deoxyadenosine
    (ii) Adenine, N-glycosidic linkage
    (iii) DNA
15. 5'– OH, Phosphodiester
16. *A*–Hydrogen bonds, *B*–Purine base, *C*–Pentose (deoxyribose) sugar, *D*–5' end, *E*–Pyrimidine
17. (i) True, (ii) True
18. Sugar and phosphates
19. Hydroxyl group (–OH)
20. Hydrogen bonds
21. A-2, B-1, C-4, D-3
22. A + T and G + C are constant and equal to 1 in all tested organisms.
23. True
24. 3'-A T G G C T G G T 5'
25. Identical
26. Antiparallel polarity.
27. According to the Chargaff's rule, A = T and G = C, hence, the G + C content is 30%.
    So, A + T = 70 %
    Thus, percentage of thymine is 35 %.
28. Two H - bonds
29. In a *ds* DNA, adenine always forms two hydrogen bonds with thymine from opposite strand and guanine forms three hydrogen bonds with cytosine and *vice-versa*, which places a purine opposite to pyrimidine base.
30. *A*–3.4 nm, *B*–0.34 nm, *C*-minor groove and *D*–20 Å.
31. Right
32. Pitch of the helix.
33. True
34. Watson and Crick
35. The genetic information flows from (ii) DNA → (iii) RNA → (i) Protein.
36. *A*–Replication, *B*–Transcription, *C*–Translation
37. Reverse transcription
38. This large length of DNA is packaged inside the nucleus by undergoing compaction. The negative charged DNA wrap around positively charged histones.
39. False. In prokaryotes, the DNA is not scattered throughout the cell, but held with some proteins around central region.
40. Nucleoid
41. 3 (DNA, histone core protein, linker H1).
42. (i) *A*–DNA , *B*–H1 histone, *C*–Histone octamer
    (ii) Nucleosome
43. Beads on a string.
44. Net positive
45. These are basic amino acids present in histone proteins.
46. (i) Histone octamer
    (ii) Chromatin
47. Chromatin
48. Beads on string, electron microscope
49. *A*–Repeating units of chromatin, *B*–Additional proteins in packaging of chromatin, *C*–Loosely packed and stains light, *D*–Heterochromatin
50. Heterochromatin
51. Transforming principle
52. *Streptococcus pneumoniae*
53. This indicates that the S-strain of this bacterium has a mucous (polysaccharide) coat.
54. (i) *A*–Mice die, *B*–Mice live
    (ii) The S-strain is virulent as a result the mice die from pneumonia, but mice infected with R–strain does not develop pneumonia so the mice live.
55. In this case, the mice live because the S-strain has lost its virulence due to heat.
56. *X* – The mice die
57. True
58. Genetic material
59. Oswald Avery, Colin MacLeod and Maclyn McCarty.
60. Nucleic acid
61. RNases
62. Bacteriophage
63. Sulphur and phosphorus
64. (i) *A*–Infection by phage, *B*–Blending, *C*–Centrifugation to separate the virus particles.
    (ii) Radioactivity was detected in ^{32}P labelled DNA cells
    (iii) It was concluded that DNA is the genetic material not the protein.
65. A-3, B -1, C-4, D-2
66. RNA
67. Replication
68. Both DNA and RNA
69. (i) False. Chemically and structurally stable nucleic acid is DNA.
    (ii) True
70. RNA

71. 2'-OH, reactive, labile

72. Thymine

73. (i), Viruses having RNA genome and shorter lifespan can mutate and evolve at a faster rate.

74. Transmission of genetic information.

75. Inheritance is odd one out. Rest others are processes evolved around RNA.

76. Semiconservative replication

77. (i) True
(ii) Watson and Crick .
(iii) Separation, complementary, template.

78. *Escherichia coli*

79. Matthew Meselson and Franklin Stahl.

80. (i) The figure represents Meselson and Stahl's experiment demonstrating semiconservative replication of DNA.
(ii) At *A*, *E. coli* cells are grown in medium containing heavy isotope of nitrogen.
(iii) CsCl gradient.

81. 75%

82. Radioactive thymidine.

83. Replication takes place in 5′ - 3′ direction on both the strands.

84. DNA polymerase-III

85. *d*NTPs (Deoxyribonucleoside triphosphates)

86. Their efficiency indicates the rate of polymerisation of a large number of nucleotides in a very short time.

87. 4.6×10^6 bp, 38 minutes.

88. 2000 bp per sceond.

89. A mistake during replication would result in mutations.

90. Terminal phosphates

91. A-4, B-3, C-1, D-5, E-2

92. Replication fork

93. False. DNA replication begins at a specific site called replication origin (*ori*) in DNA.

94. Lagging

95. *A*–Leading strand, *B*–Lagging strand

96. Okazaki fragments

97. DNA ligase

98. (i) Replication fork.
(ii) *A* is the single-stranded binding proteins which keep the unwounded strands in an extended form.
(iii) False. *B* is helicase which catalyses unwinding of parental double helix.
(iv) *C* is a primer which provides a starting point for replication.
(v) *D* is DNA polymerase.

99. This is because the vectors provide the origin of replication essential for their propagation within the host genome.

100. Restriction enzyme is odd one out. All others are a part of DNA replication process.

101. S-phase

102. This condition of failure in cell division after DNA replication results in polyploidy.

103. Transcription

104. Uracil

105. False. In transcription, only a segment of DNA and only one of the strands is copied into RNA.

106. Different sequences

107. Formation of double-stranded RNA would prevent RNA from being translated into proteins, thus the exercise of transcription becomes futile.

108. False. Transcription occurs unidirectionally in 5'- 3' direction.

109. 4 (Promoter, structural gene, RNA coding gene and terminator).

110. Coding strand

111. RNA sequence will be UAA - GGG - CCG - CGU - ACG - UAC - GUA - CGU.

112. *A*–Promoter, *B*–Terminator, *C*–Template strand, D–Coding strand.

113. (i) Promoter is located upstream of the structural gene.
(ii) Terminator is located downstream of the structural gene.

114. Promoter

115. Terminator

116. Pribnow box

117. Gene

118. Cistron

119. Eukaryotes

120. Genes in the eukaryotes are sptit means that genes have interrupted coding sequences.

121. Exons

122. Exons

123. A-4, B-2, C-1, D-3.

124. *r*RNA

125. RNA polymerase

126. Promoter region

127. Single

128. False. RNA polymerase does not require primer and lacks the ability to repair mistakes in RNA.

129. 2 (core enzyme and holoenzyme)

130. The correct sequence will be c (chain initiation), a (chain elongation) and b (chain termination).

131. *A*– Formation of transcription bubble, *B*–Phosphodiester bond.

132. Sigma factor

133. Ribonucleotide triphosphates

134. ρ

135. (ii), The *m*RNA in bacteria does not require any processing to become active, also transcription and translation take place in the same compartment.

136. Cytoplasm

137. (i) RNA polymerase-I
(ii) RNA polymerase-III
(iii) RNA polymerase-II

138. An additional complexity present in eukaryotic transcription is that the primary transcripts contain both the exons and the introns which make it non-functional.

139. Transformation is odd one out. Rest others are post-transcriptional modifications undergone by primary transcripts.

140. False. Introns are excised and hydrolysed during post-transcriptional processing.

141. Capping and Tailing. Addition of a methyl guanosine triphosphate to 5' end, cleavage and polyadenylation to create 3'end.

142. (i) Transcription in eukaryotes
(ii) Primary transcript
(iii) Splicing
(iv) Capping, adenylate, polyadenylation

143. False. The fully processed *hn*RNA is called *m*RNA which is transported out from the nucleus.

144. 7-methylguanosine

145. Genetic code

146. *A*–Proposed that genetic code is a triplet, *B*–Har Gobind Khorana, *C*–Cell free system for protein synthesis, *D*–Severo Ochoa.

147. (i) Triplet
(ii) Universal
(iii) Commaless/Contiguous
(iv) Degenerate

148. True

149. UAA

150. GUG is the odd one out because it is an initiation codon. Rest others are termination codon.

151. Last

152. Termination, non-sense

153. Amino acid coded by the given sequence will be Met, Leu, Phe, Lys, Cys, Gly, Phe.

154. The *m*RNA chain that can be translated completely is, AUG UAU UUC UGC CUC UAG.

155. CCA is odd one out as it codes for amino acid proline. Rest others are codes for leucine.

156. GGC, GGA and GGG.

157. Mutation

158. True

159. Silent

160. (ii), An almost completely altered sequence.

161. Reading frame of triplet codon

162. Point

163. Beta globin, valine

164. DNA - CAC/GTG, RNA - GUG, Amino acid - valine

165. A common result of insertion and deletion processes is change in the reading frame.

166. False. An adapter molecule is *t*RNA.

167. Francis Crick

168. (i) *A*–Aminoacyl synthetase binding loop or D-loop, *B*–Acceptor stem, *C*–Ribosomal binding loop or TψC loop, *D*–Anticodon loop, *E*–Variable loop
(ii) AGU

169. (i) There was a need of an adapter molecule because amino acids cannot read the code.
(ii) A = 5′, B = Anticodon, C = Codon

170. Robert Holley (1965)

171. Translation

172. False. The order and sequence of amino acids are defined by the sequence of bases in the *m*RNA.

173. *m*RNA

174. (i) Translation, (ii) *A*–Leucine and *B*–Serine.

175. The correct sequence is as follows

Activation of amino acids, Initiation, Elongation/ Chain formation, Termination and Modifications.

176. Binds to the *t*RNA holding the growing polypeptide chain.

177. 9 (Initiation factors, polyribosomes, elongation factors, translocase, GTPase, *m*RNA, *t*RNA, aminoacyl *t*RNA synthetase, amino acids).

178. 'X' is charging of *t*RNA or aminoacylation of *t*RNA.

179. Untranslated Region

180. False. The untranslated regions are required for efficient translation.

181. Ribozyme

182. 40S and 60S

183. *A*–Initiator *t*RNA, *B*–Amino acid-*t* RNA, complexes binds sequentially to *m*RNA codons *C*–Sequential translation of amino acids into polypeptide sequence, *D*–Binding of release factors.

184. Transcriptional level, processing level, transport of *m*RNA from nucleus to the cytoplasm and translational level.

185. Constitutive

186. *A*–Beta-galactosidase

187. (ii), If bacteria do not have lactose, around them to be utilised for energy source, they would no longer require the enzyme beta-galactosidase, since it hydrolyses lactose.

188. True

189. RNA, accessory

190. Operon

191. True

192. Polycistronic

193. (i) Off
(ii) Repressor
(iii) *X* or repressor binds to operator and prevents RNA polymerase from transcribing the operon.
(iv) Jacob, Monod

194. (i) False. Lactose is present in the medium.
(ii) Promoter gene
(iii) Operator gene
(iv) Regulator gene

195. A-2, B-3, C-1.

196. *A*-*i* gene, *B* - RNA polymerase gets access to the promoter and transcription proceeds.

197. True

198. Negative

199. True

200. Genetic engineering made isolation and cloning of any piece of DNA easier and faster.

201. Human Genome Project

202. 1990

203. HGP was called a megaproject because of its large magnitude and reqirements as it involved sequencing storge, retrieval and analysis of each and every base pair in the human genome.

204. 3 billion

205. False. HGP was closely associated with bioinformatics.

206. 20,000-25,000 genes

207. False. The information was stored in databases.

208. 13

209. The US Department of Energy and the National Institute of Health.

210. 9 billion US dollars

211. Improving manpower for data analysis was not an objective of HGP.

212. 7 (Bacteria, yeast, *Caenorhabditis elegans, Drosophila*, rice, *Arabidopsis*, chimpanzee).

213. Expressed sequence tags and sequence annotation.

214. Sequence annotation.

215. *A*–Cloning in a suitable host using specialised vectors, *B*–PCR, *C*–Fragments were sequenced using automated DNA sequencers, *D*–Alignment of these sequences.

216. Chain termination method

217. Bacteria and yeast

218. Chromosome 1

219. (ii), on polymorphism of restriction endonuclease recognition sites and satellite DNA.

220. Dystrophin

221. Less than 2%

222. False. 99.9% nucleotide bases are similar among all peoples.

223. Repetitive

224. Y-chromosome

225. Single Nucleotide Polymorphisms

226. DNA fingerprinting

227. Satellite DNA that have repeating short stretches of DNA with high degree of polymorphism.

228. No, these sequences normally do not code for any proteins.

229. (i), DNA fingerprinting is the basis of paternity testing in case of disputes because satellite DNA that contributes to polymorphisms are inheritable from parents to children.

230. Polymorphisms

231. True

232. Germ cells

233. DNA polymorphism

234. Alec Jeffreys

235. Variable Number of Tandem Repeat

236. Minisatellites

237. The steps in sequence are (iv) → (ii) → (i)→ (iii) → (v) → (vi)

238. Restriction endonuclease

239. False. VNTR varies in size from 0.1 to 20 kb.

240. Monozygotic twins

241. Polymerase Chain Reaction (PCR)

242. Suspect is *B* because it matches (AR) the most with DNA found at the crime scene.

CHAPTER SEVEN

EVOLUTION

Topic 1 : Origin of Life and Evolution of Life Forms

1. Name the branch of science which deals with the study of history of life forms on earth.

____

2. Changes in the flora and fauna on earth in past times can be understood by studying evolution of earth, stars and universe.

True ☐ False ☐

3. Stellar distances are measured in _______________(kilometers/light years).

4. The stars we see today emitted light _______________ years back.

5. Match the following columns.

Column I	Column II
A. Universe	1. 4.5 billion years
B. Earth	2. 4 billion years
C. First life form on earth	3. 20 billion years

A.__________ B.__________ C. __________

6. Name the theory which explains the origin of universe as a singular huge explosion unimaginable in physical terms.

____

7. Arrange the following steps in the sequence of their occurrence during universe formation.

I. Formation of simplest elements, i.e. hydrogen and helium.
II. Cataclysmic eruption of hot, energy rich subatomic particles.
III. Universe expansion and lowered temperature.
IV. Formation of galaxies.
V. Condensation of gases.

____

8. Who proposed the Big-Bang theory?

____

9. The atmosphere on early earth was rich in oxides and hydrogen.

True ☐ False ☐

10. How many of the listed components covered the surface of early earth?

Water vapours, Methane, Carbon dioxide, Ammonia

____

11. Complete the following flowchart by filling blanks *A, B* and *C* depicting the various reactions that might occurred on early earth.

Water vapours —Ⓐ→ Escape hydrogen + Ⓑ

↓

Water + CO_2 + Ammonia + Ⓒ

A.______________ B.______________ C.______________

12. The __________ were formed on earth when __________ layer was cooled and water vapours fell as rain to fill all the depressions.

13. Match the following columns.

Column I (Theories of origin of life)		Column II (Postulates)
A. Theory of panspermia	1.	Life came out of decaying and rotting matter like straw and mud.
B. Theory of spontaneous generation	2.	World has passed through many ages and at the end of each age, all living beings died.
C. Theory of catastrophism	3.	Units of life or spores were transferred to different planets including earth.

A.______________ B.______________ C.______________

14. Louis Pasteur demonstrated that life comes from pre-existing life and thus, dismissed the theory of biogenesis.

(*Rewrite the statement by changing the underlined words.*)

____

15. Refer to the diagram showing Louis Pasteur experiment and answer the questions that follows.

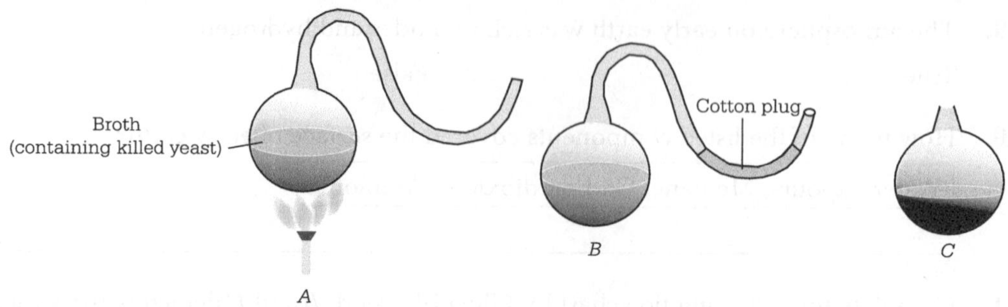

(i) In which of the following flask, microorganisms were likely to grow faster?

____

(ii) What was the significance of curved/swan-necked flask?

____

16. Choose the odd one out.

Francisco Reddi, Lazzaro Spallanzani, Louis Pasteur, Von Helmont

____

17. Oparin and Haldane proposed that first form of life could have come from ______________________________(Pre-existing, living, inorganic molecules/ Pre-existing, non-living, organic molecules).

18. Formation of diverse organic molecules from inorganic constituents occurred during ____________________ evolution.

19. According to Oparin and Haldane, chemical evolution occurred after the biological evolution on earth.

True ☐ False ☐

20. How many of the listed conditions were found on the early earth?

Low temperature, Volcanic storms, Reducing atmosphere, CH_4 and NH_3 in atmosphere

____

21. Name the scientist who created early earth conditions in a laboratory scale.

____

22. Refer to the given diagram and answer the questions that follows.

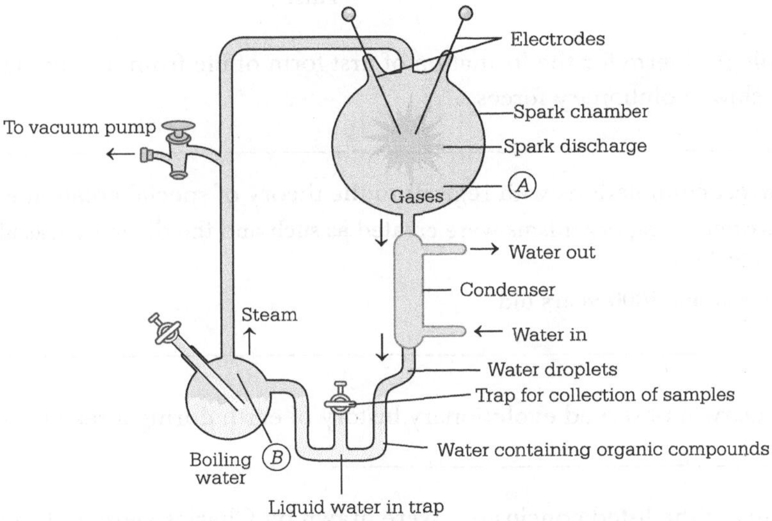

(i) Name the three gases in flask labelled as *A*.____________.

(ii) What was the temperature inside flask *B*, which formed water vapours (H_2O)?

____

(iii) What was formed and observed by Miller in the flask *A*?

____

(iv) In the similar setup, what was formed and observed by other scientists in the flask *A*?

____

23. The compounds formed in Miller's experiment were similar to those in earth's crust. (*Rewrite the statement by changing the underlined word.*)

____

24. The first non-cellular life forms must have been originated __________ years ago in the form of giant molecules like ____________________.

25. The first cellular form of life is believed to have originated about ____________ (2000/200) million years ago (mya).

26. Complete the following analogy given below.

First non-cellular forms : Giant molecules : : First cellular forms : ____________

27. All life forms are believed to be originated on land only.

True ☐ False ☐

28. Give biological term for the formation of first form of life from non-living molecules through slow evolutionary forces.

____

29. An incorrect connotations with regards to the theory of special creation is

(i) All existing living organisms were created as such and the diversity was always the same since creation.

(ii) Earth is about 3000 years old.

____

30. Charles Darwin observed evolutionary history of earth during a sea voyage in a ship called ______________ .

31. How many of the listed conclusions were drawn by Charles Darwin during his voyage around world?

Similarities of living forms with past forms, Extinction of life forms, Gradual evolution of life forms, Formation of new life forms, No variations in a population.

____

32. According to Darwin, reproductive fitness helps organisms to better fit in an environment leave more progeny than others.

True ☐ False ☐

33. Give technical term for the phenomenon which acts as mechanism of evolution (according to Darwin).

____

34. ______________, a naturalist who worked in Malay Archipelago and concluded that all existing life forms share ______________ ancestors which were present in ______________ periods in the history of earth.

35. The geological history of earth does not coincide with the biological history of earth.

True ☐ False ☐

Topic 2 : Evidences of Evolution

36. Give biological term for the remains of hard parts of life forms found in rocks.

____

37. The rocks in which fossils are found, are called ____________________.

38. In the cross-section of earth's crust, the primitive most fossil would be found in ________________ (lower sediments/upper sediments).

39. Refer to the family tree of dinosaurs given below and answer the questions that follows.

(i) Name the organisms labelled as *A*, *B* and *C*.

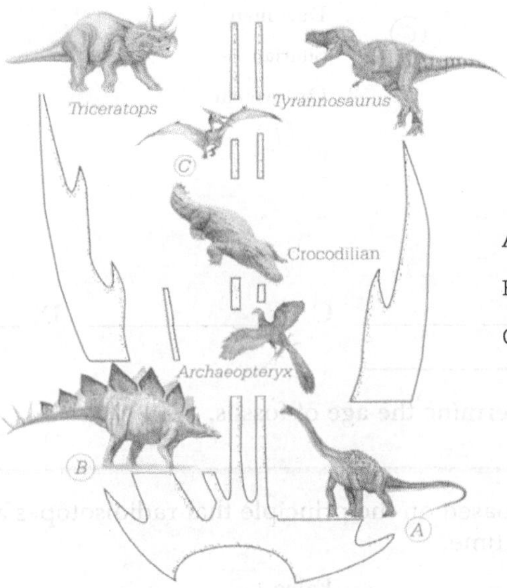

A. ____________________

B. ____________________

C. ____________________

(ii) The given family tree is showing extinct animals only.

Yes ☐ No ☐

(iii) Name any two modern day counterpart organisms of dinosaurs.

____

40. The study of fossils in different sedimentary layers indicates the <u>Permian period</u>. (*Rewrite the statement by changing the underlined words.*)

____

41. Arrange the following in the order of decreasing magnitude in geological time scale.

I. Periods, II. Eras, III. Epochs, IV. Eons

____

42. Complete the flowchart representing the geological time scale.

Eon	Era	Period
(A)	(B)	Quaternary
		Neogene
		(D)
	Mesozoic	Cretaceous
		(E)
		Triassic
	(C)	Permian
		Carboniferous
		Devonian
		Silurian
		Ordovician
		(F)
Proterozoic	–	–
Archean	–	–
Hadean	–	–

A.__________ B.__________ C. __________ D. __________

E.__________ F. __________

43. Name any two methods to determine the age of fossils.

____

44. Radioactive dating method is based on the principle that radioisotopes are stable and accumulate over the course of time.

True ☐ False ☐

45. Complete the following analogy with respect to the conversion of a radioisotope into its non-radioactive form.

Carbon^{14} : Nitrogen^{14} : : Potassium^{40} : __________

46. Who proposed the embryological support for evolution?

____

47. Answer the following questions.

(i) The law which states 'ontogeny recapitulates phylogeny' is known as __________.

(ii) The proposal about embryological support for evolution was disapproved by

____

48. How many of the listed examples support embryological evidences of evolution?

Vestigial gill slits in human embryo, Ciliated sperms in bryophytes and pteridophytes, Neoteny in axolotl larvae, Reptilian characters in *Archaeopteryx*

____

49. Comparative anatomy and morphological studies focus on whether common ancestors among organisms were shared or not.

True ☐ False ☐

50. Refer to the given diagram and answer the questions that follows.

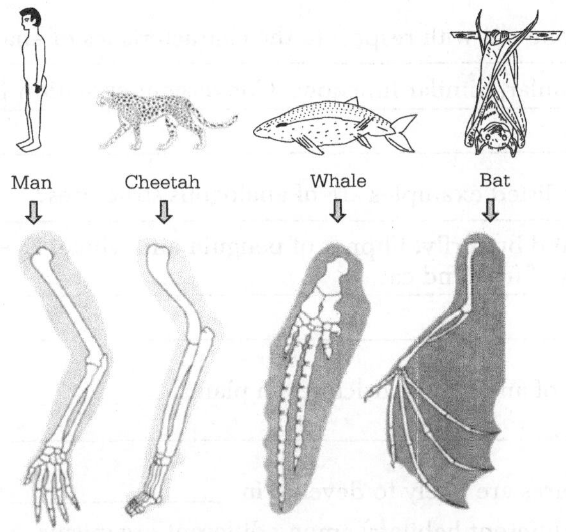

(i) Give the common biological term for these structures.

____

(ii) The forelimbs of different animals shown in diagram are ____________ similar, but ____________ dissimilar.

(iii) Name the type of evolution shown by these structures.

____

51. Refer to the diagram and identify whether these structures are homologous or vestigial.

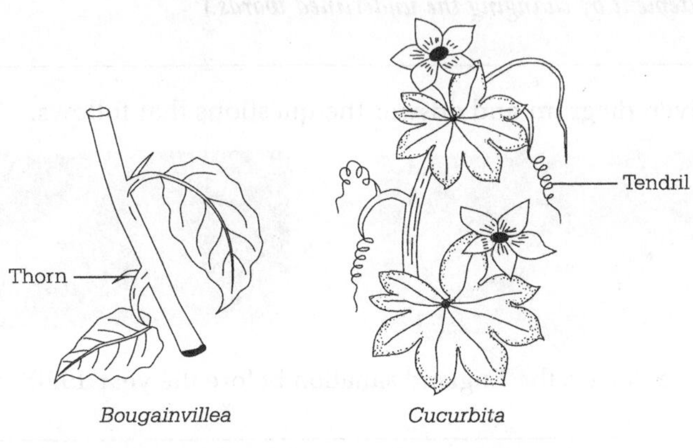

____

52. Vertebrate heart or brain show convergent evolution.
(*Rewrite the statement by changing the underlined words.*)

____

53. Complete the following analogy given below.
Homology : Common ancestry :: ______________ : Different ancestry.

54. Choose the odd one out with respect to the characteristics of analogous structures.

Anatomically similar, Similar functions, Convergent evolution

____

55. How many of the listed examples are of analogous structures?

Wings of birds and butterfly, Flipper of penguin and whale, Eyes of octopus and mammals, Brain of frog and cat

____

56. Give an example of analogous structures in plants.

____

57. Analogous structures are likely to develop in ______________________________ (similar habitats/different habitats) among different organisms.

58. Similarities in proteins and genes performing a given function indicates __________ ancestry. This biochemical similarity is related to the ______________ similarity among diverse organisms.

59. Complete the following analogy given below.
Vestigial organs : Remnants of functional organs in ancestor :: ____________ : Sudden reappearance of ancestral characters.

60. Industrial melanism in moths of England between 1850-1920 support evolution by convergent evolution.
(*Rewrite the statement by changing the underlined words.*)

____

61. Refer to the given diagram and answer the questions that follows.

A *B*

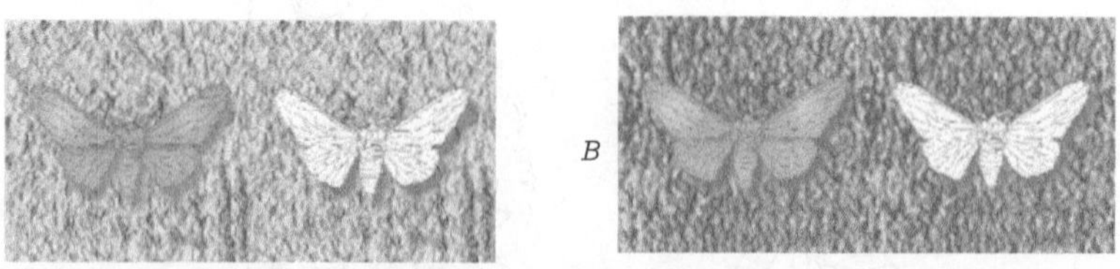

(i) Which picture depicts the England situation before the year 1920 ?

____

(ii) What does the light coloured background represent in diagram *A* ?

____

(iii) The condition shown in diagram *A* favoured the survival of ________________________ (both moths/light-winged moths).

(iv) The population of ________________ winged moths was more in diagram *B* because they were able to ________________ themselves from predators.

(v) The given diagram depicts industrial melanism where the term melanism is applicable to ____________________ (lichens/dark-winged moths).

62. Give an example of evolution by anthropogenic actions (other than industrial melanism).

____

63. Evolution by anthropogenic actions occurs in centuries.

True ☐ False ☐

64. Evolution is not a direct process in the sense of determinism. It is a ________________ process based on chance events in nature and chance of ______________ in organisms.

Topic 3 : Adaptive Radiation

65. Name the islands on which Darwin observed the birds shown in the diagram given below.

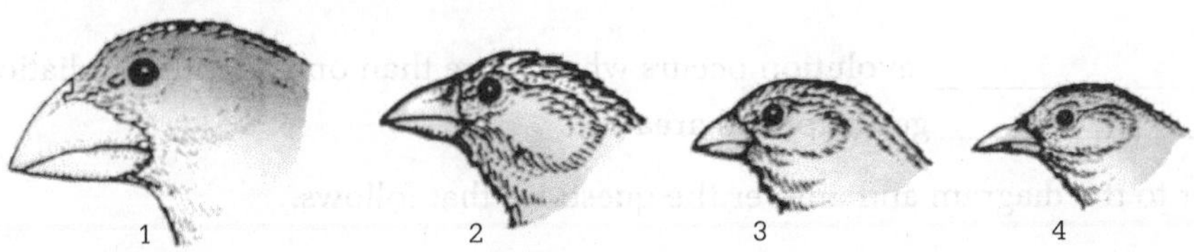

____

66. The varieties of Darwin finches had ______________________ (altered beaks/different eye colours).

67. The original (ancestral) Darwin finches was insectivorous.
(*Rewrite the statement by changing the underlined word.*)

____

68. The seed eating ancestor finches of Galapagos island gave rise to vegetarian finches only.

True ☐ False ☐

69. Give biological term for 'the process of evolution of different species in a given geographical area starting from a point and radiating to other habitats.'

____

70. Complete the following analogy given below.

Industrial melanism : Natural selection :: Australian marsupials : ________________

71. Refer to the given diagram and identify the type of evolution shown in it.

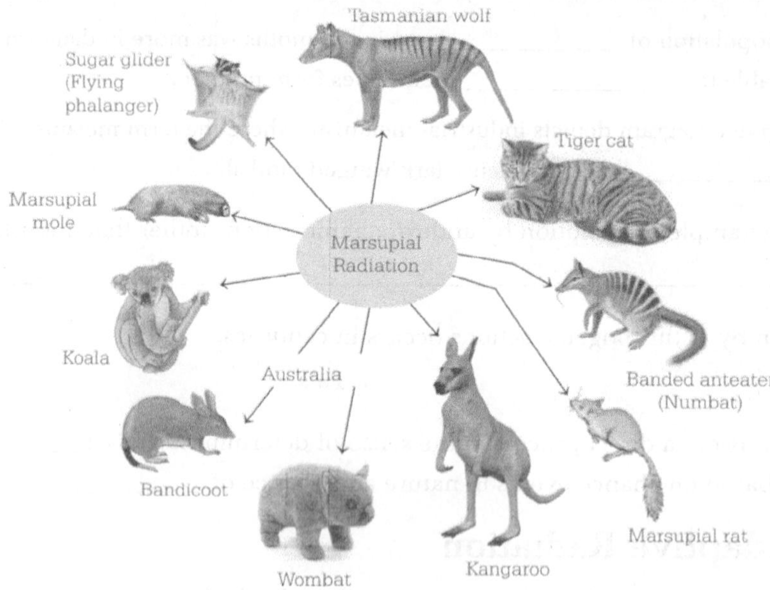

____

72. ______________ evolution occurs when more than one adaptive radiations occur in ______________ geographical area.

73. Refer to the diagram and answer the questions that follows.

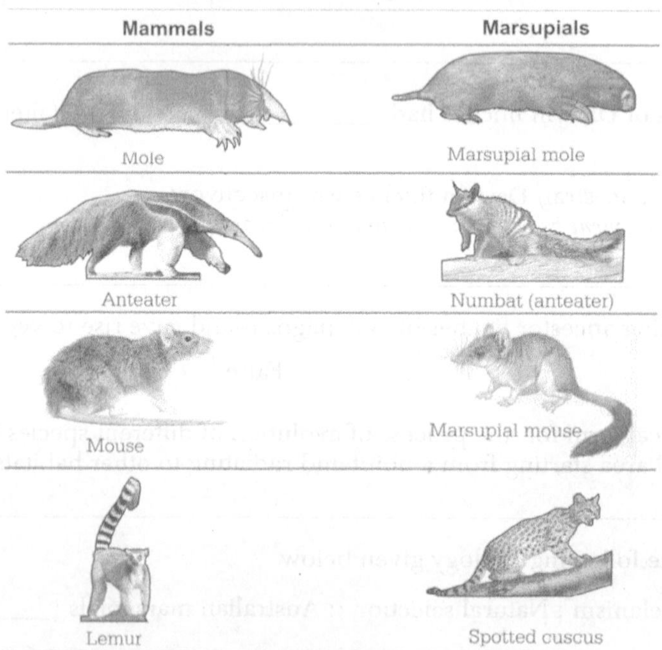

(i) The animals on the left hand side are ________________________________

(ii) Give the collective term for the animals shown on the right hand side.

____

(iii) Both the categories of animals show ________________ evolution.

Topic 4 : Biological Evolution and its Mechanism

74. The presence of cellular life forms with different metabolic capability on earth represents the evolution by ______________________________ .

75. Refer to the diagrams showing two populations under different conditions.

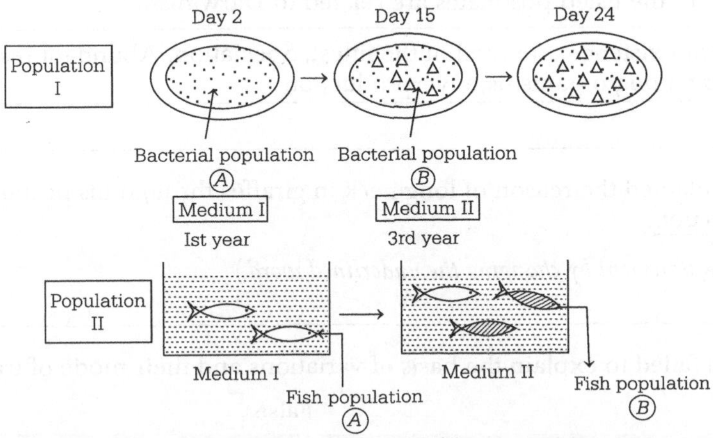

How many of the listed conclusions can be drawn from the above diagrams?

(i) The rate of appearance of new life forms is not related to life cycle of organism.

(ii) Population *B* in both cases (I and II) is better than *A* in terms of fitness.

____

76. Adaptive ability is inherited and ______________ is the end result of the ability to adapt and get selected by nature.

77. Name the two key concepts of Darwinian Theory of Evolution.

____

78. Match the following columns.

Column I (Scientists)	Column II (Associated works)
A. Lamarck	1. Nature selects organisms with high reproductive rate
B. Darwin	2. An essay on population
C. Thomas R. Malthus	3. Use and disuse of organs

A.______________ B.______________ C. ______________

79. Arrange the following steps of Lamarck's theory of use and disuse of organs, in the order of his description.
 I. Long neck in giraffes. II. Leaf foraging on tall trees.
 III. Acquired character of elongated neck in generations.

____

80. Lamarck's theory was published in his book called

____

81. Who discarded the Lamarck's theory by showing that tail-less mice produce mice with tail only?

____

82. How many of the listed postulates are related to Darwinism?

Struggle for existence, Survival of the fittest, Speciation, Abundant resources, Variations among organisms, Overproduction

____

83. Darwin explained the reason of long neck in giraffes through his postulate of <u>overproduction</u>.

(*Rewrite the statement by changing the underlined word.*)

____

84. Darwinism failed to explain the basis of variations and their mode of transmission.

True ☐ False ☐

85. The idea of mutation was put forth by ______________ based on his work on ______________.

86. Complete the following analogy given below.

Darwin : ______________ : : Hugo de Vries : Mutation causes evolution.

87. In contrast of Darwinian variations, the mutations are ______________ (random and directionless/ small and directional).

88. Give the biological term for single step large mutation which causes speciation.

____

Topic 5 : Hardy-Weinberg Principle

89. Name the principle which helps to find out the frequency of occurrence of alleles of a gene or a locus.

____

90. According to Hardy-Weinberg principle, the allele frequency in a population is not stable from generation to generation.

True ☐ False ☐

91. Give biological term for the total genes and their alleles in a population.

____

92. The total genes in a population are said to be in genetic equilibrium when sum total of all allelic frequencies is ____________(one/zero).

93. Hardy-Weinberg principle is calculated by the binominal expansion of $\underline{p^2 + q^2}$. (*Rewrite the statement by changing the underlined word.*)

____

94. Hardy-Weinberg principle is calculated by the equation $p^2 + 2pq + q^2 = 1$. Complete the given table based on the representation of each component of equation.

p	Frequency of dominant allele
p^2	(A)
q	(B)
q^2	Percentage of homozygous recessive individuals
$2pq$	(C)

A.____________ B.____________ C. ____________

95. The genotypes AA, Aa, aa have frequencies 0.5, 0.25 and 0.25, respectively. What will be the frequency of A and a?

____

96. The frequency of allele p is 0.6 and q is 0.4 at a particular locus. What would be the frequency of heterozygotes in this population?

____

97. The difference between the expected and obtained frequency in Hardy-Weinberg law indicate the extent of ______________________________.

98. All the listed factors are required in population exhibiting Hardy-Weinberg principle, except one. Identify it.

Small population size, Random mating, No gene flow, No genetic drift, No selection and mutation

____

99. In a population, Hardy-Weinberg equilibrium is disrupted due to gene migration.

Yes ☐ No ☐

100. Match the following columns.

	Column I (Factors)		Column II (Explanations)
A.	Mutation	1.	Changes in population's frequencies due to chance alone.
B.	Gene flow	2.	Differences in survival and reproduction among variant individuals.
C.	Natural selection	3.	Changes in allele frequency due to immigration and emigration.
D.	Genetic drift	4.	Sources of new alleles.

A.____________ B.____________ C.____________ D. ____________

101. The migration of a section of population *A* to a new population *B* changes the gene frequency in ______________ (*A* only/*A* and *B*).

102. Name the phenomenon in which change in allele frequencies drift the original population and form new sample of population having different species.

____

103. Drastic reduction in a population even if its organisms do not move from place to place is known as ______________.

104. Give biological term for the process of formation of new phenotypes through pre-existing advantageous mutations. ______________

105. ______________ is a process in which heritable variations enable better survival and reproduction rate.

106. How many listed factors result in changed frequency of genes and alleles in future generation?

Mutations, Recombination, Gene flow, Genetic drift

____

107. Refer to the diagram and answer the questions that follows.

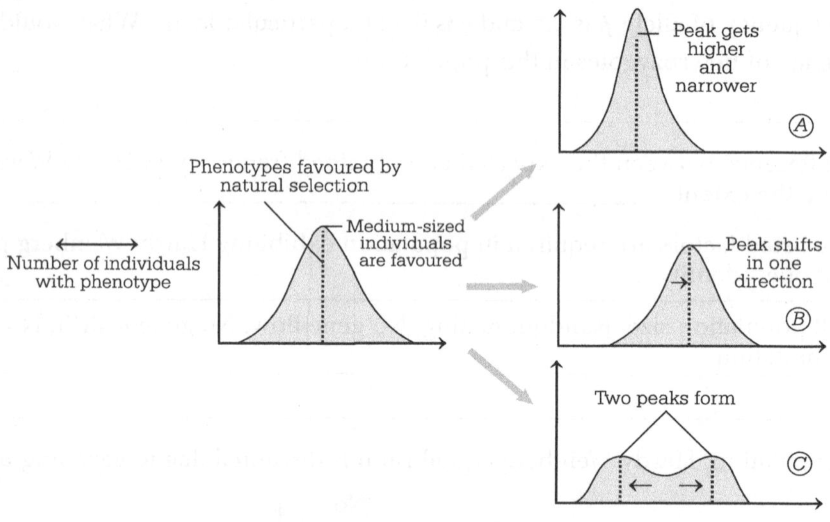

(i) Name the type of natural selection shown in *A*, *B* and *C*.

A.______________ B.______________ C.______________

(ii) The type of selection shown in *A* does not promote evolutionary change but maintain phenotypic stability within the population.

True ☐ False ☐

(iii) The *B* and *C* types of natural selection occur due to ______________ ______________ (stable environment and no selection/selective pressure under changing environment).

108. Complete the following analogy given below.
Industrial melanism : Directional selection :: Sickle-cell anaemia : ______________

Topic 6 : A Brief Account of Evolution

109. When did the first cellular forms of life appeared on earth?

____

110. On early earth, the cells with membranous envelope had the ability to release ______________. The reaction could have been similar to the ________ reaction in photosynthesis where ________ is splitted with the help of solar energy.

111. Arrange the following organisms in the order of parts with regarding the recent evolvement on earth.

I. Sea weeds, II. Jawless fish, III. Invertebrates

____

112. Name/Give examples for the following.

(i) First organisms that invaded land. ______________

(ii) Fish that could move on land and in water. ______________

(iii) Ancestors of modern day frogs and salamanders. ______________

(iv) Modern day reptilian descendents of amphibians. ______________

113. Match the following columns.

	Column I		Column II
A.	Pteridophytes	1.	Fish-like reptiles in 200 mya
B.	*Ichthyosaurs*	2.	Biggest dinosaur with dagger-like teeth
C.	*Tyrannosaurus*	3.	Formed coal deposits

A.__________ B.__________ C.__________

114. About <u>150 mya</u>, the dinosaurs suddenly disappeared from earth.
(*Rewrite the statement by changing the underlined word.*)

____

115. The first mammals were like bats.

True ☐ False ☐

116. Mammal's fossils are ______________ sized. They were viviparous, and took over the earth when ______________ came down.

117. The South American mammals resembling horse, bear, rabbit, etc., were over-ridden by North American fauna due to ________________ (recombination/continental drift).

118. Pouched mammals of Australia survived even after continental drift due to the lack of ________________ from any other mammal.

119. Name any two mammals that live wholly in water.

____

120. Given below is the sketch of evolution of plants forms through geological periods, examine it and answer the questions that follows.

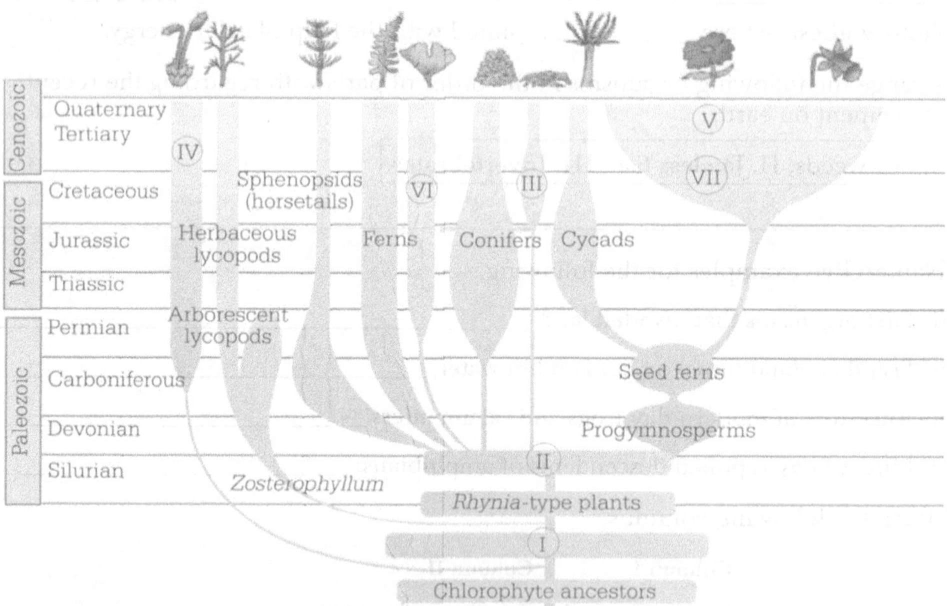

(i) Mention the name of plant types at I and II position.

____

(ii) Bryophytes are likely to be found at position (IV/VII).

____

(iii) Complete the following analogy given below.

VI : Ginkgos : : III : ________________

(iv) Name the type of plants likely to be found at position V.

____

121. Refer to the representative evolutionary history of vertebrates through geological period and answer the questions that follows.

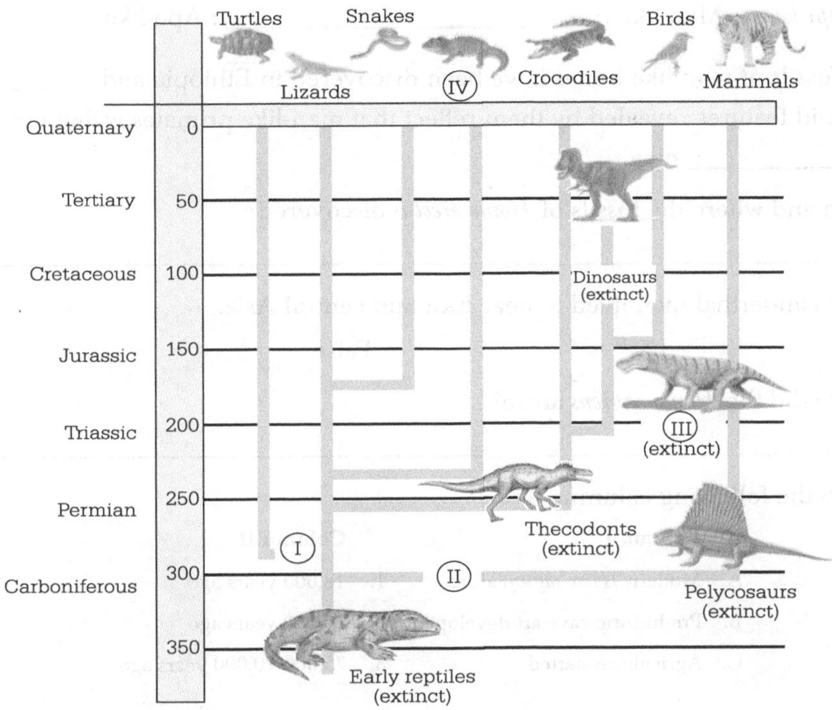

(i) Name the animals labelled as I and II. ____________________

(ii) Tuataras would be found at position (III/IV) ____________________

(iii) ____________________ are not shown in the diagram before early reptiles because there are no specimens of these left with us.

Topic 7 : Origin and Evolution of Man

122. Complete the flowchart depicting human evolution.

Dryopithecus ⟶ (A) ⟶ *Australopithecus* ⟶ (B)
↓
Homo sapiens ⟵ Cro-Magnon ⟵ (C) ⟵ *Homo erectus*

A.__________ B.__________ C. __________

123. Complete the table given below (*A-E*).

Names	Ages of appearance	Cranial capacities	Important features
Ramapithecus	15 mya	–	(A)
Australopithecus	5 mya	(B)	Hunted with stone weapons and ate fruits
Homo habilis	2 mya	(C)	First human-like being the hominid, did not eat meat
Homo erectus	1.5 mya	900 cc	(D)
Neanderthal man	100,000-40,000 years	(E)	Used hides, burried their dead

A.__________ B.__________ C. __________ D. __________

E.__________

124. Complete the following analogy given below.
Ramapithecus : Man like :: ________________________ : Ape-like

125. Few fossils of man like bones have been discovered in Ethiopia and ____________. The hominid features revealed by them reflect that man-like primates walked in _______ about ______________ mya.

126. When and where the fossils of *Homo erectus* discovered?

____

127. The Neanderthal man lived in near East and central Asia.
True ☐ False ☐

128. Where did the *Homo sapiens* arose?

____

129. Match the following columns.

Column I	Column II
A. Modern *Homo sapiens* arose	1. 18,000 years ago
B. Pre-historic cave art developed	2. 10,000 years ago
C. Agriculture started	3. 75,000-10,000 years ago

A.______________ B.______________ C. ______________

130. Complete the following analogy given below.
Cro-Magnon man : 1650 cc :: Living modern man : ________________________

131. Refer to the diagram showing comparison of the skulls and answer the questions that follows.

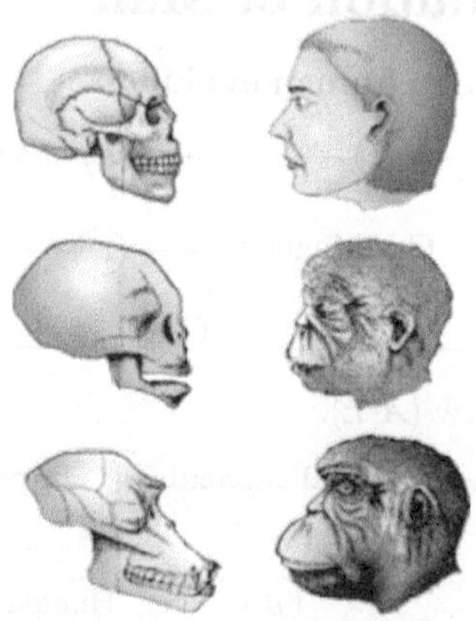

(i) The skull II and III belong to which animals?

____

(ii) Which one is likely to possess greater cranial capacity? (I/III)

____

Answers

1. Evolutionary biology.
2. True
3. Light years
4. Million
5. A-3, B-1, C-2
6. Big-Bang theory
7. II $\rightarrow$ III $\rightarrow$ I $\rightarrow$ V $\rightarrow$ IV
8. Georges Lemaitre in 1920.
9. False. There was no atmosphere on the early earth.
10. 4 (All of them)
11. *A*-UV rays, *B*-oxygen *C*-Methane.
12. Oceans, ozone
13. A-3, B-1, C-2
14. Theory of spontaneous generation.
15. (i) Flask *C*, because broth is exposed to environment.
    (ii) To trap the germ particles in the curve of the neck preventing them from reaching the broth contamination.
16. Von Helmont is odd one out, as he was a supporter of theory of spontaneous generation. While the other scientists supported theory of biogenesis.
17. Pre-existing, non-living, organic molecules.
18. Chemical
19. False. According to Oparin and Haldane, chemical evolution occurred before the biological evolution on earth.
20. 3 (All of them except low temperature). Early earth had high temperature instead of low temperature.
21. SL Miller
22. (i) Methane (CH_4), ammonia (NH_3), hydrogen (H_2).
    (ii) 800°C
    (iii) Amino acids
    (iv) Sugar, nitrogen bases, pigments and fats.
23. Meteorites
24. 3 billion, RNA, protein or polysaccharide
25. 2000
26. Single cells
27. False. All life forms are believed to be originated in water environment only.
28. Abiogenesis
29. (ii) is incorrect because according to theory of special creations, earth is about 4000 years old.
30. HMS Beagle
31. 4 (All of them except no variations in a population) Darwin observed built in variations in populations which enable them to survive in natural conditions.
32. True
33. Natural selection
34. Alfred Wallace, common, different
35. False. The geological history of earth closely correlates with the biological history of earth.
36. Fossils
37. Sedimentary rocks
38. Lower sediments
39. (i) *A-Brachiosaurus*
    B-Stegosaurus
    C-Pteranodon
    (ii) No. The diagram shows both extinct dinosaurs and their living modern day counterparts like crocodiles and birds.
    (iii) Birds and crocodiles.
40. Geological period
41. IV $\rightarrow$ II $\rightarrow$ I $\rightarrow$ III
42. *A*–Phanerozoic, *B*–Cenozoic, *C*–Paleozoic, *D*–Paleogene, *E*–Jurassic, *F*–Cambrian.
43. Radioactive dating method and Electron spin resonance method.
44. False. Radioisotopes are unstable and they decay slowly over a period of the time and thus, used to determine the age of fossils.
45. $Argon^{40}$
46. Ernst Haeckel
47. (i) Biogenetic law
    (ii) Karl Ernst von Baer
48. 3 (All of them except reptilian characters in *Archaeopteryx*). Reptilian characters in *Archaeopteryx* is an example of palaeontological evidence (connecting link fossil).
49. True
50. (i) Homologous organs/structures
    (ii) Anatomically, functionally
    (iii) Divergent evolution
51. Homologous structures
52. Divergent evolution
53. Analogy
54. Anatomically similar is odd one out, as analogous structures are not similar anatomically.
55. 3 (All of them except brain of frog and cat). Brain of frog and cat is an example of homologous structures.
56. Potato (stem modification) and sweet potato (root modification)
57. Similar habitats
58. Common, structural
59. Atavism
60. Natural selection
61. (i) Diagram *A*, i.e. before industrialisation.
    (ii) Light coloured background represents lichens growing on tree trunks.
    (iii) Light-winged moths
    (iv) Dark, camouflage
    (v) Dark-winged (melanised) moths.
62. Selection of resistant varieties due to the excessive use of herbicides or pesticides.
63. False. Evolution by anthropogenic actions usually occurs in much lesser time scale of months or years.
64. Stochastic, mutation
65. Galapagos islands
66. Altered beaks
67. Seed-eater

68. False. The original seed eating ancestor finches of Galapagos island gave rise to insectivorous and vegetarian finches.

69. Adaptive radiation

70. Adaptive radiation (other example is Darwin finches).

71. Divergent evolution

72. Convergent, isolated

73. (i) Placental mammals
(ii) Australian marsupials
(iii) Convergent

74. Natural selection

75. Only one (ii), i.e. population *B* is better than *A*.

76. Fitness

77. Branching of descent and natural selection

78. A-3, B-1, C-2

79. II → III → I

80. *Philosophie Zoologique* in 1809.

81. August Weismann

82. 5 (All of them except abundant resources). Darwin did not considered abundant resources, rather he said that resources remain limited inspite of rapid multiplication of individuals.

83. Survival of the fittest or natural selection

84. True

85. Hugo de Vries, Evening primrose

86. Heritable variations causes evolution

87. Random and directionless

88. Saltation

89. Hardy-Weinberg principle

90. False. According to Hardy-Weinberg principle, the allele frequency in a population remains stable and constant from generation to generation.

91. Gene pool

92. One

93. $(p+q)^2$

94. *A* - Percentage of homozygous dominant individuals.
B - Frequency of recessive allele
C - Percentage of heterozygous individuals.

95. $A = AA + \frac{1}{2}Aa$

$= 0.5 + \frac{1}{2}(0.25) = 0.625$

$a = 1 - A = 0.375$

96. Frequency of heterozygotes $= 2pq = 2 \times 0.6 \times 0.4 = 0.48$

97. Evolutionary changes

98. Small population size. Hardy Weinberg principle is applicable to large populations.

99. Yes

100. A-4, B-3, C-2, D-1

101. A and B

102. Founder effect

103. Bottle neck effect

104. Speciation

105. Natural selection

106. 4 (All of them)

107. (i) *A*–Stabilising
B–Directional
C–Disruptive
(ii) True
(iii) Selective pressure under changing environment.

108. Disruptive selection

109. 2000 million years ago (mya)

110. Oxygen, light, water

111. III → II → I
(500 mya) (350 mya) (320 mya)

112. (i) Plants
(ii) Coelacanth
(iii) Lobefins
(iv) Turtles, tortoises and crocodiles

113. A-3, B-1, C-2

114. 65 mya

115. False. The first mammals were like shrews.

116. Small, reptiles

117. Continental drift

118. Competition

119. Dolphins, whale, sea cows.

120. (i) I-Tracheophytes ancestors, II-Psilophyton
(ii) Position IV
(iii) Gnetales
(iv) Angiosperms

121. (i) I-Sauropsids, II-Synapsids
(ii) Position IV.
(iii) First amphibians

122. *A-Ramapithecus, B-Homo habilis, C-Homo sapiens neanderthalensis*

123. *A*-Hairy and walked like gorilla and chimpanzees, *B*-500 cc, *C*-650-800 cc, *D*-Ate meat, used stone and bone tools, *E*-1400 cc

124. *Dryopithecus*

125. Tanzania, Eastern Africa, 3-4

126. In Java, 1891

127. True

128. Africa

129. A-3, B-1, C-2

130. 1450 cc

131. (i) II-Baby chimpanzee, III-Adult chimpanzee
(ii) I, i.e. adult human has cranial capacity 1450 cc, whereas it is about 350-450 cc in chimpanzee.

CHAPTER EIGHT

HUMAN HEALTH AND DISEASE

Topic 1 : Human Health and Common Diseases In Humans

1. The state of complete physical, mental and social well-being is ____________

2. How many factors in the given list are necessary for achieving good health?

Yoga, Drugs, Balanced diet, Vaccination, Alcohols, Hygienic food and Water

____

3. Complete the flowchart given below (*A-D*).

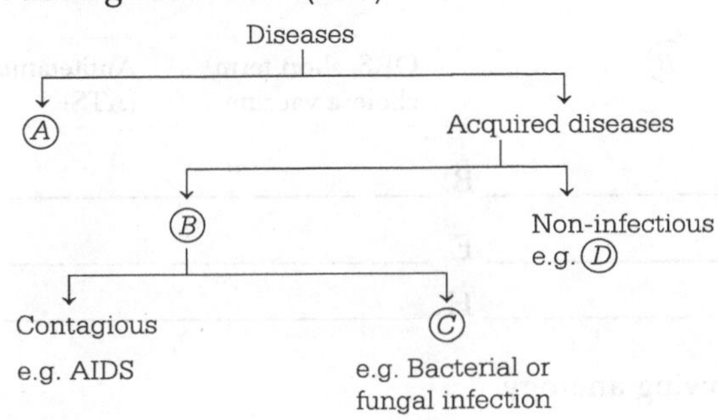

A.____________ B. ____________

C.____________ D. ____________

4. The recent pandemic caused due to Covid-19 is an example of ____________ disease.

5. Give the biological term for the disease causing organisms. ____________

6. Pathogenic organisms have the ability to adapt and resist physiological activities within the host body.

True ☐ False ☐

7. Name the causative agent of typhoid fever in human beings.

___

8. (i) The pathogenic bacterium which causes typhoid invades ___ through contaminated food and water. Its symptoms include sustained high fever, ___ and ___. Typhoid fever could be confirmed by ___ test.

(ii) The lady who continued to spread typhoid for several years through food was ___

9. Answer the following questions in context of pneumonia.

(i) Two causative agents ___

(ii) Infected organ(s) ___

(iii) Symptoms (Any 4) ___

(iv) Transmission ___

10. Complete the table depicting various bacterial diseases in humans.

	Tuberculosis	Cholera	Tetanus	Pertussis
Causative agent	*Mycobacterium tuberculosis*	(C)	(E)	(G)
Transmission	Droplet infection	(D)	Spores released from animal faeces enter through wounds	Airborne bacteria infect respiratory tract
Symptoms	(A)	Diarrhoea, vomiting	(F)	Whooping cough, vomiting, fever, etc.
Treatment	(B)	ORS, short term cholera vaccine	Antitetanus Serum (ATS)	(H)

A.___ B. ___ C. ___

D.___ E. ___ F. ___

G.___ H. ___

11. Complete the following analogy.

Plague : *Yersinia pestis* : : Leprosy : ___

12. Leprosy is transmitted to man through infected rat flea, *Xenopsylla cheopis*. (*Rewrite the statement by changing the underlined word.*)

___

13. Read the following statements and identify the bacterial disease.

I. Its pathogen was first identified by GA Hansen in 1870.

II. It is characterised by scaly and modulated skin, deformed fingers, hypopigmented skin, etc.

III. Treated by rifampicin and dofazimine drugs, BCG vaccine, etc.

___

14. The causative agent of diphtheria is ____________________

15. Diphtheria vaccine, i.e. ____________ (DPT/BCG) is given in first six weeks of birth.

16. Give the full form of the following terms.

(i) BCG ____________________

(ii) DPT ____________________

(iii) NLEP ____________________

17. Name the STD caused by the bacterium *Treponema pallidum.* ____________

18. (i) Name the group of viruses which cause the most infectious human ailments, i.e. the common cold. ____________________

(ii) Mention any two symptoms of common cold. ____________

19. Choose the odd one out with respect to the symptoms of disease caused by rhinoviruses.

Nasal congestion, Lung fibrosis, Sore throat, Nasal discharge

____

20. Name the common group of viruses to which SARS, MERS and Covid-19 belong.

____

21. Match the following columns.

Column I (Viral diseases)	Column II (Causative agents)
A. Chickenpox	1. Rhabdovirus
B. Rabies	2. Enterovirus
C. Poliomyelitis	3. Paramyxovirus
D. Influenza	4. Zoster virus
E. Dengue	5. *Flavivirus*

A.____________ B.____________ C. ____________ D. ____________

E.____________

22. Salk vaccine is given orally to prevent poliomyelitis.

True ☐ False ☐

23. *Varicella* vaccine helps to prevent dengue.

True ☐ False ☐

24. The vector of dengue fever is *Aedes aegypti* mosquito.

True ☐ False ☐

25. Pick the odd virus, based on the mode of transmission.

Hepatitis-B virus, Chikungunya virus, Ebola virus, Human immunodeficiency virus

____

26. *Plasmodium* parasite that causes ____________ (plague/malaria) in neither a virus nor 9 bacterium.

27. Malignant malaria is caused by <u>*Plasmodium malariae.*</u>
(*Rewrite the statement by changing the underlined word.*)

____

28. Complete the flowchart given below.

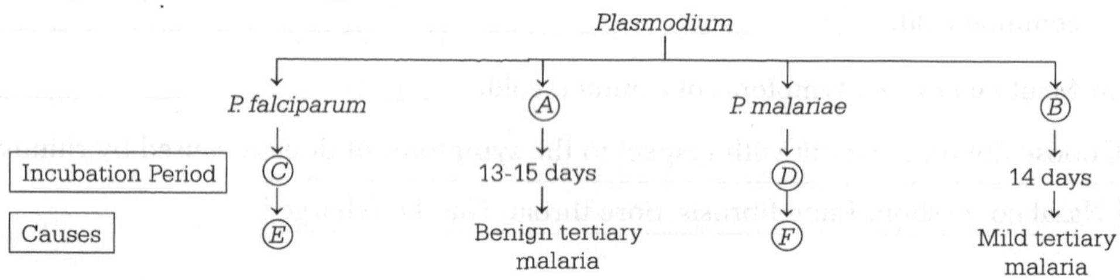

A. ________________ B. ________________

C. ________________ D. ________________

E. ________________ F. ________________

29. Complete the following analogy given below regarding malarial pathogen.

Intermediate host : Human : : Definitive host : ____________________

30. (i) Both sexual and asexual life cycle of *Plasmodium* occur in human.

True ☐ False ☐

(ii) *Anopheles* acts as host as well as vector of malarial parasite.

True ☐ False ☐

31. Refer to the given diagram and identify *A*, which is being released by female *Anopheles*.

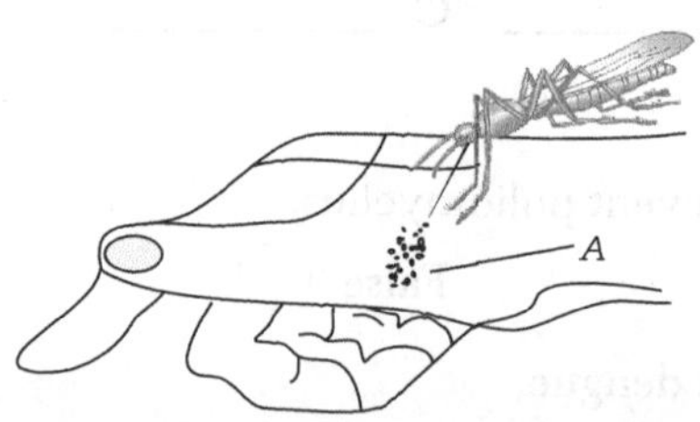

A. ____________________

32. The liver of person infected with malaria contains ________________ stage of parasite. After reproducing ________________, the parasite burst liver cell and release ________________ into the blood. The latter invade RBCs in blood.

33. Arrange the following stages of malarial parasite in the sequence of their occurrence in human RBCs.

I. Schizont containing schuffner's dots.
II. Amoeboid stage containing haemozoin granules.
III. Signet ring stage.
IV. Gametocytes

____

34. Recurring high fever and chills during malaria is caused due to the release of merozoites from red blood cells.
(*Rewrite the statement by changing the underlined word.*)

____

35. Fertilisation and development of malarial parasite occur in the ______________ (stomach/salivary gland) of mosquito.

36. Identify the structure *A* that is found in mosquito and the stage of malarial parasite *B* shown in the diagram given below.

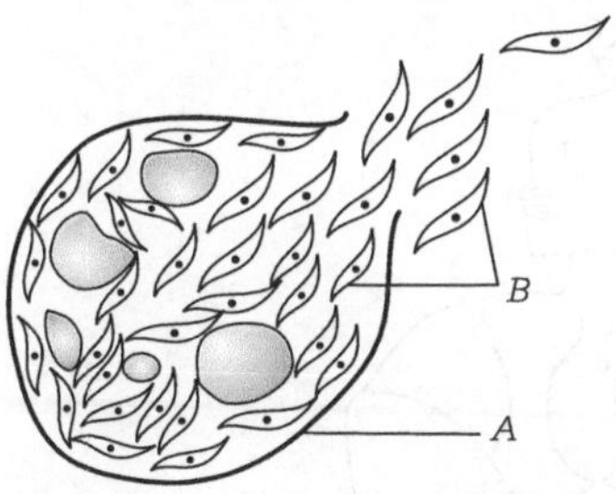

A. ______________________

B. ______________________

37. Chloroquine is the most commonly used drug to treat malaria.

True ☐ False ☐

38. Identify the protozoan parasite based on its characteristics given below.

I. Parasitised large intestine of human and causes amoebiasis.
II. Transmitted through faecal oral route by houseflies.
III. Causes abdominal pain, excess mucus and blood in stools, etc.

____

39. Match the following columns.

Column I (Protozoan diseases)	Column II (Causative agents)
A. Sleeping sickness	1. *G. lamblia*
B. Giardiasis	2. *Trypanosoma gambiense*
C. Chagas disease	3. *Leishmania donovani*
D. Kala-azar	4. *Trypanosoma cruzi*

A. ____________ B. ____________ C. ____________ D. ____________

40. How many of the protozoan diseases in the given list, transmit through faecal oral route?

Chagas disease, Giardiasis, Sleeping sickness, Amoebiasis

____

41. Name the transmitting agents of protozoan diseases listed in questions.

(i) Trypanosomiasis ____________________

(ii) Leishmaniasis ____________________

(iii) Chagas disease ____________________

42. Ascariasis is caused due to an ____________________ parasite, *Ascaris.*

43. Mention any two symptoms of ascariasis.

____

44. The medical condition shown in the given diagram is caused due to a helminthic infection. Consider the diagram and answer the questions that follows.

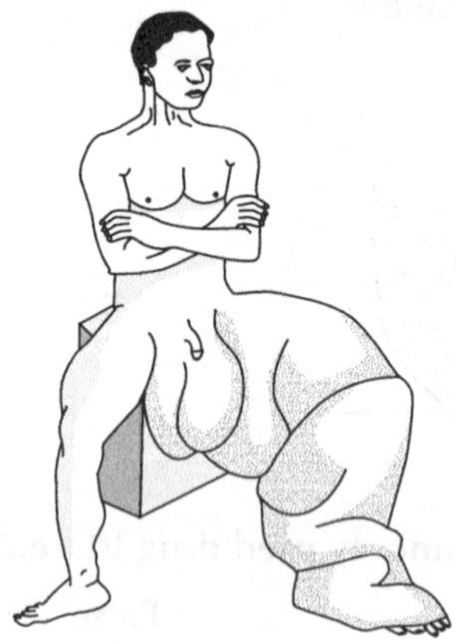

(i) Name the disease. ____________________

(ii) Name the pathogenic organism. ____________________

(iii) What is the mode of transmission of this disease?

____

45. Taeniasis is caused by the infection of *Taenia solium*, which is found in cattle.

True ☐ False ☐

46. Complete the following analogy given below.

Wuchereria : Microfilariae : : *Taenia* : ____________________

47. How many of the given fungal species cause ringworms in human?

Microsporum, Trichophyton, Epidermophyton

____

48. The pathogen of ringworm grows in the area of human body which is ______________________ (hot and moist/cold and dry).

49. Athlete's foot is an infection caused by the fungal genus – ______________________ .

50. All the listed fungal genus do not invade the deeper tissues of the body, except

Microsporum, Trichophyton, Blastomyces

____

51. Match the following columns.

	Column I (Preventive measures)		Column II (Types of disease)
A.	Consumption of clean water and food	1.	Pneumonia, common cold
B.	Avoid close contact with infected person	2.	Malaria, filariasis
C.	Eliminate breeding places of mosquitoes	3.	Typhoid, amoebiasis, ascariasis

A.____________ B.____________ C. ____________

52. (i) Introduction of *Gambusia* fish in ponds help to <u>prevent</u> <u>pneumonia</u>.
(*Rewrite the statement by changing the underlined word.*)

____

(ii) <u>Tetanus</u> have been eradicated completely through the use of vaccines and immunisation.
(*Rewrite the statement by changing the underlined word.*)

____

Topic 2 : Immunity

53. The ability of the body to fight the disease causing organisms is called

______________________ .

54. The non-specific type of defence, that is present at the time of birth is ____________

55. Innate immunity function by acting as a barrier to the entry of foreign agent into body.

True ☐ False ☐

56. Choose the odd one out with respect to the type of innate immunity barrier.

Stomach acid, Mucus coating of respiratory tract, Saliva, Tears

____

57. Complete the following analogy given below.

Skin : Physical barrier : : ______________________ : Cellular barrier

58. Monocytes, macrophages and natural killer cell in the blood form ______ (cellular/physiological) barriers in body.

59. Give the scientific term for the proteins secreted by virus-infected cells.

______

60. Name the type of barrier which protects body against viral infections.

______

61. Match the following columns.

	Column I (Types of innate immunity barrier)		Column II (Modes of functioning)
A.	Cellular barriers	1.	Prevent microbial growth
B.	Physiological barriers	2.	Fight against viral infections
C.	Cytokine barriers	3.	Trap microbes
D.	Physical barriers	4.	Phagocytose and destroy microbes

A. ______ B. ______

C. ______ D. ______

62. The pathogen specific immunity, that is characterised by memory is called ______ immunity.

63. An attack by a pathogen for the first time produces high intensity primary response.

True ☐ False ☐

64. Anamnetic response is alternatively known as ______

65. Refer to the figure given below and answer the following questions.

(i) Name the parts labelled as *A*, *B* and *C*.

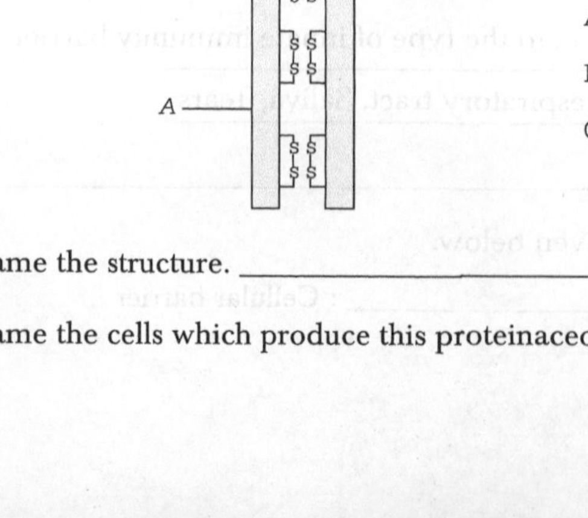

A. ______

B. ______

C. ______

(ii) Name the structure. ______

(iii) Name the cells which produce this proteinaceous structure. ______

66. The structure of antibody molecule is represented as ______________ (H_4L_2 / H_2L_2).

67. The upper and lower portion of each chain of antibody possess ______________ and __________ regions, respectively.

68. Complete the table depicting various characteristics of different antibodies.

Types of antibody	Percentage of serum	Functions
IgG	(A)	Stimulate phagocytes and complement system.
IgM	5-10%	(B)
(C)	15-20%	Protect exposed surface of body and thus found in seromucus secretions.
(D)	Less than 1%	Mediate hypersensitivity reactions and allergic response.
IgD	(E)	Activate B - cells to secrete antibodies.

A. ______________________ B. ______________________ C. ______________________

D. ______________________ E. ______________________

69. <u>IgA</u> is the only antibody that can cross placenta from mother to child and provides immune protection to newborns..
(*Rewrite the statement by changing the underlined word.*)

____

70. Complete the following analogy given below.
Smallest antibody : IgG : : Largest antibody : ______________________

71. Antibodies formed by identical B-cells are called polyclonal antibodies.
True ☐ False ☐

72. Antibodies or immunoglobulins are synthesised by animals to combat
____

73. Give technical term for different antigen-antibody interactions describe below.
(i) An antibody binds to more than one antigen to form a large insoluble complex. ________.
(ii) The process by which an antibody makes the antigen susceptible for phagocytosis.
____
(iii) An antibody binds with soluble antigen so as make it insoluble and promote its phagocytosis.
____

74. Complete the flowchart depicting the types of immune response by different lymphocytes.

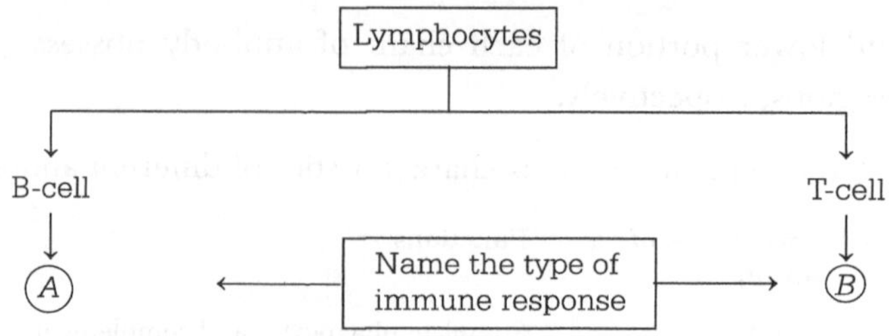

A. ____________________ B. ____________________

75. Which of the following graft is likely to be rejected by body? (Autograft/Xenograft)

____

76. Humoral immune response is responsible for the graft rejection in humans.
(*Rewrite the statement by changing the underlined word.*)

____

77. How many of the listed procedures can be used as the preventive measures of graft rejection?

Tissue matching, Blood-group matching, Immunosuppression, Exposure of bone marrow to X-rays

____

78. Read the following statements and identify the type of immunity.
(i) Antibodies are produced in host body upon exposure to an antigen.
(ii) It is slow but long lasting.

____

79. Active immunity is generated in response to a living pathogen (antigen) only.

True ☐ False ☐

80. Give an example of artificially acquired active immunity.

____

81. In passive immunity, ____________________ are given to protect the body against foreign agents.

82. Choose the odd one out.

Colostrum, Antitoxin, TAB vaccine

____

83. Name the type of antibody found in colostrum which provides natural passive immunity.

____

84. Antisera contain ______________ antibodies to provide artificial passive immunity. This immunity provides immediate relief and is ______________ lived.

85. The purpose of vaccination is to ____________________________ (kill the pathogens /generate memory cells).

86. During vaccination, antigens are introduced in the body which later help to elicit anamnestic response on subsequent infection.

True ☐ False ☐

87. The treatment given for tetanus and snakebite is based on the principle of <u>active</u> immunity.
(*Rewrite the statement by changing the underlined word.*)

____

88. The vaccines produced using RDT usually contain ______________________________

89. Give the biological term for the exaggerated response of immune system to certain antigens present in the environment. ______________________________

90. Name any two allergens which are naturally found in environment.

____

91. Complete the flowchart depicting the process of allergic reaction in human body.

Allergen $\xrightarrow{\text{Bind to}}$ (A) $\xrightarrow{\text{Bind to}}$ Mast cells $\xrightarrow{\text{Release}}$ (B) $\xrightarrow{\text{Causes}}$ Sneezing, Coughing, etc.

A.______________________________ B. ______________________________

92. Adrenaline help to quickly reduce the symptoms of allergy.

True ☐ False ☐

93. Rheumatoid arthritis is an example of ____________________________ because the body attacks self-cells.

94. Choose the odd one out with respect to the autoimmune disease.

Diabetes mellitus, Multiple sclerosis, SCID, Lupus, Myasthenia gravis

____

95. Give the collective term used for bone marrow and thymus.

____

96. (i) Immature lymphocytes differentiate into antigen sensitive lymphocytes in ____________________________ (primary/secondary) lymphoid organs.

(ii) ______________ is a lobed lymphoid organ near the heart and beneath breastbone.

97. How many secondary lymphoid organs are given below?

Spleen, Thymus, Tonsils, Lymph nodes, Peyer's patches, Appendix

____

98. Refer to the given figure and answer the following questions.
   (i) Name the structure labelled as *A*, *B* and *C*.

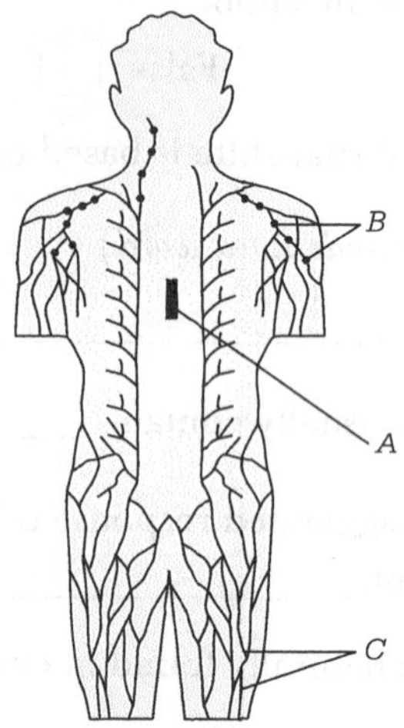

A. ____________________

B. ____________________

C. ____________________

   (ii) Which structure provides the sites for interaction of lymphocytes with the antigen (*A*/*B*)? __________

   (iii) _______ (*A*/*C*) provides micro-environment for the development and maturation of T-lymphocytes.

99. Match the following columns.

Column I (Lymphoid organs)	Column II (Characteristics)
A. Bone marrow	1. Trap antigens present in tissue fluid
B. Thymus	2. Site of production of all lymphocytes
C. Spleen	3. Reduces in size with advancing age
D. Lymph nodes	4. Trap blood borne microorganisms

A. __________ B. __________ C. __________ D. __________

100. (i) Lymphoid tissue located within the lining of major tracts like respiratory, digestive, etc., is called ________________. It constitutes ________ per cent of lymphoid tissue in human body.

   (ii) ________________ is the large bean-shaped lymphoid organ which contains lymphocytes and ________________ .

Topic 3 : AIDS

101. AIDS is a congenital disease.

True ☐ False ☐

102. AIDS is associated with the _____________ of immune system. It is known to cause group of symptoms therefore it is a _______________ . AIDS was first reported in the year _____________ .

103. Name the virus which causes AIDS.

____

104. AIDS causing virus belongs to the group of viruses called adenovirus. (*Rewrite the statement by changing the underlined word.*)

____

105. The genetic material in HIV is ____________________ (naked DNA/enveloped RNA).

106. Choose the odd one out with respect to the transmission of HIV infection.

Sexual contact with multiple partners, Taking drugs intravenously, Repeated blood transfusions, Sharing articles with infected person

____

107. Refer to the given diagram showing replication of HIV in human and answer the following questions.

(i) Name the structures labelled as *A* and *B*.

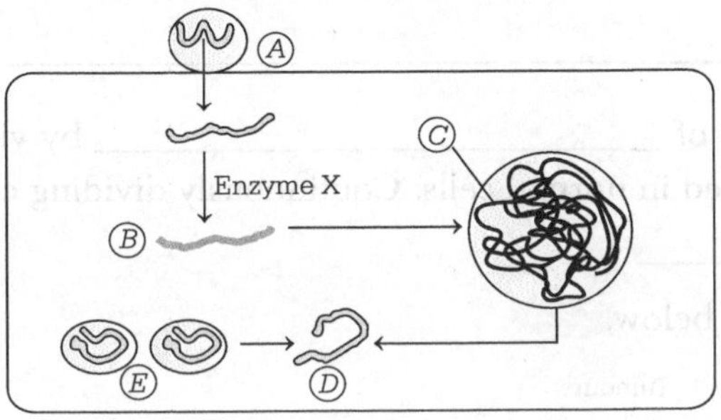

A. ____________________

B. ____________________

(ii) Name the enzyme *X* involved in the production of *B*. ____________________

(iii) The formations of structure *B* occurs in the ____________________ (macrophages/nucleus) of host cells.

(iv) Identify the structures labelled as *C* and *D*.

C. ____________________ D. ____________________

(v) The infected host cells do not survive after viruses are being replicated and released from them.

True ☐ False ☐

108. Helper T-lymphocytes acts as HIV factory in the body of person infected with AIDS.
(*Rewrite the statement by changing the underlined words.*)

____

109. Name the cells which progressively decreases in number in the body of AIDS infected person. ______________________________

110. Various symptoms of AIDS like weight loss, fever, etc., appear when

____

(Viral RNA replicate in macrophages/Helper T-lymphocytes decreases).

111. A widely used diagnostic test for AIDS is ______________________

112. Expand the following.

(i) NACO ______________________________

(ii) NGO ______________________________

113. How many of the listed measures given below help to prevent AIDS?

Use of disposable syringes, Advocating safe sex, Use of condoms during intercourse, Isolation of AIDS infected person

____

Topic 4 : Cancer

114. Give the biological term for the oncogenic transformation of cells in human body.

____

115. Cancer cells lack the property of ______________________ by virtue of which uncontrolled growth is inhibited in normal cells. Continuously dividing cells form masses of cells called ______________ .

116. Complete the flowchart given below.

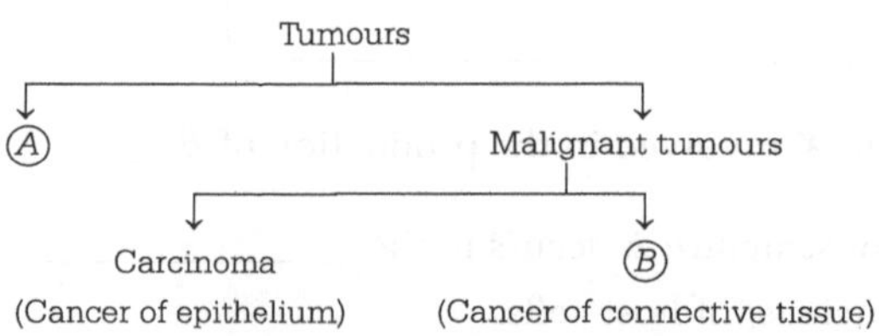

A.______________ B.______________

117. Read the following statements and identify the type of cancer.

(i) Proliferating mass of neoplastic cells.
(ii) Starve the normal cells by competing for vital nutrients.
(iii) Show metastasis.

____

118. Match the following columns.

	Column I (Forms of cancer)		Column II (Characteristics)
A.	Adenoma	1.	Spread of cancer to new sites or organs.
B.	Metastasis	2.	Cancer of pigment cells of skin.
C.	Benign tumour	3.	Cancer of glands.
D.	Melanoma	4.	Remain confined to original location.

A. ____________ B. ____________ C. ____________ D. ____________

119. Choose the odd one out with respect to the type of carcinogen.

X-rays, Gamma rays, UV-rays

____

120. Tobacco smoke causes lung cancer as it contain ____________ carcinogens.

121. Give the biological term for cancer causing viruses. ____________

122. Viral oncogenes are the type of ____________ (physical/biological) carcinogens.

123. Name the genes in normal cells which, when activated, could lead to oncogenic transformation of the cells. ____________

124. Complete the following analogy.

Cancer of internal organs : Computed Tomography : : ____________ : Blood and bone marrow test.

125. Match the following columns.

	Column I (Cancer detection techniques)		Column II (Procedures)
A.	Biopsy	1.	X-rays generate 3D image of organs.
B.	CT scan	2.	Non-ionising radiations detect pathological changes in living tissue.
C.	MRI	3.	Piece of suspected tissue is cut into thin sections.
D.	Histopathological studies	4.	Suspected tissues are observed under microscope.

A. ____________ B. ____________ C. ____________ D. ____________

126. Some genes predispose an individual to certain cancers and these can be detected by using the techniques of molecular biology.

True ☐ False ☐

127. In immunotherapy, tumour cells are irradiated lethally without affecting the normal tissues surrounding them.
(*Rewrite the statement by changing the underlined word.*)

____

128. Name the biological response modifiers which activate immune system and destroy tumour.

____

129. Side effects like hair loss and anaemia in a cancer patient are most likely to be associated with _______________ (chemotherapy/biopsy).

Topic 5 : Drugs and Alcohol Abuse

130. The biological substances used to modify or explore physiological or pathological state of body are called _______________

131. Name any three commonly abused drugs.

____

132. All the drugs are obtained from fungi.

True ☐ False ☐

133. Opioid drugs bind to their specific receptors which are found in CNS and _______________ (respiratory tract/gastrointestinal tract).

134. Read the following characteristics and identify the name of opioid drug.

(i) White, odourless, bitter crystalline compound.

(ii) Commonly called smack.

(iii) Obtained by acetylation of morphine.

____

135. Chemically heroin is triacetylmorphine.
(*Rewrite the statement by changing the underlined word.*)

____

136. Identify the correct structure of morphine.

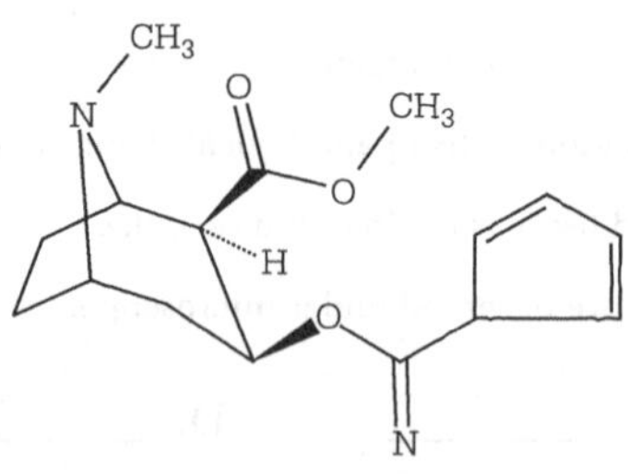

(i)

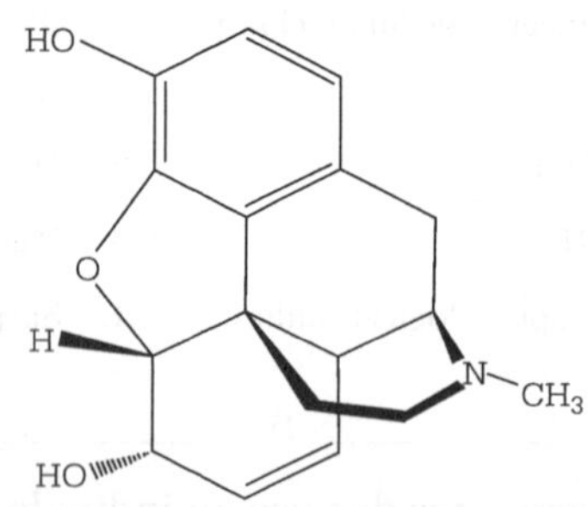

(ii)

____

137. Refer to the given figure and answer the following questions.

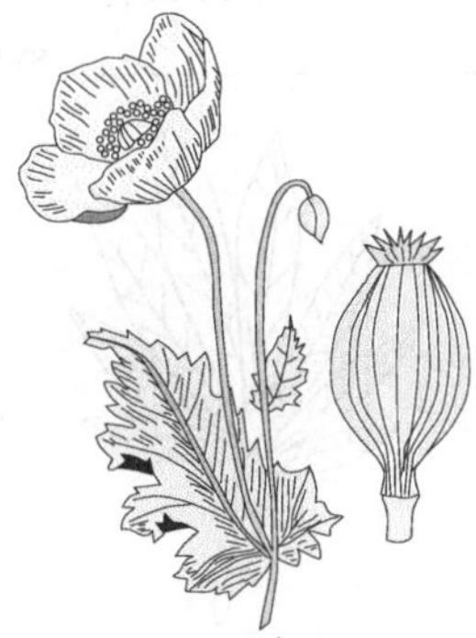

(i) Name the plant (both common and scientific names).

____

(ii) Name the drug obtained from this plant. ______________

(iii) Which part of this plant is used to obtain the drug? ______________

138. Choose the odd one out with respect to the mode of taking heroin.

Injection, Snorting, Orally

____

139. Heroin is a depressant.

True ☐ False ☐

140. Complete the following analogy.

Natural opiate : Heroin : : Synthetic opiate : ______________

141. The cannabinoid receptors are principally found in ______________

142. Name the plant and the part from which the cannabinoids are obtained.

____

143. Give the common name for the drugs given below.

Marijuana, Hashish, Charas, Ganja

____

144. Identify the group labelled as *A* in the structure of cannabinoid molecule given below.

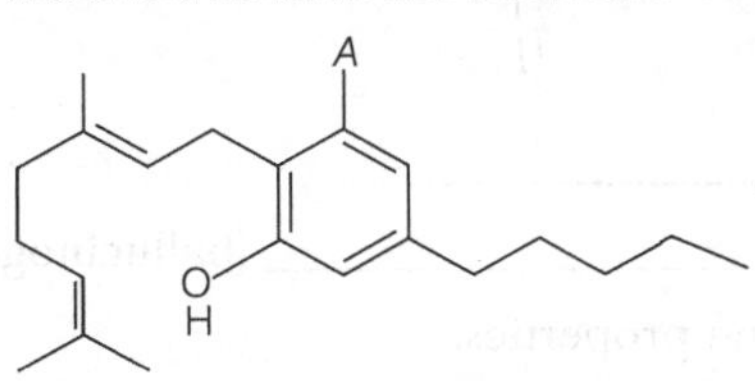

A. ______________

145. The diagram given below shows the leaves of ______________________ which affect ______________________ system of the body.

146. Complete the following analogy.

Morphine : *Papaver somniferum* : : Cocaine : ______________________

147. Match the following columns.

	Column I		Column II
A.	Coke	1.	Sedative and painkiller
B.	Morphine	2.	Inhaled or orally ingested
C.	Cannabinoid	3.	Snorted

A. ______________ B. ______________

C. ______________

148. Cocaine is commonly called coke or crack and it interferes with the transport of neurotransmitter ______________________. It ______________________ the CNS and produce a sense of euphoria.

149. Refer to the given diagram and answer the following questions.

(i) Name the plant. ______________________

(ii) The given plant has ______________________ (hallucinogenic/sedative) properties.

(iii) Name the two plants with same properties.

____

150. How many of the listed drugs are used as medicines to help patients cope with mental illness?

Barbiturates, Amphetamines, Hashish, Benzodiazepines

____

151. Sports persons usually abuse cocaine.
(*Rewrite the statement by changing the underlined word.*)

____

152. Expand LSD. ______________________________

153. Answer the questions with respect to tobacco.

(i) The alkaloid found in tabacco is ______________

(ii) Tobacco smoking can raise blood pressure and heart rate due to the release of hormones ______________ and ______________ from adrenal gland.

(iii) Tobacco smoking increases ______________ content in blood and reduces haembound oxygen.

154. Adolescents are motivated toward drug and alcohol use due to ______________ (excitement/biological demands).

155. All the listed factors might encourage youngsters to consume drugs and alcohol, except.

Peer pressure, Social media, Stress, Unsupportive family, Mental illness

____

156. Give technical term for the psychological attachment with drugs and alcohol for certain effects like euphoria.

____

157. Read the following statements and identify the technical term which describes the consequence of alcohol and drug abuse.

(i) Tendency of body to manifest withdrawl symptoms if regular dose of drugs/alcohol is abruptly discontinued.

(ii) Ignoring of all social norms to satiate self needs.

____

158. Choose the odd one out with respect to the symptoms of withdrawl syndrome.

Anxiety, Sweating, Euphoria, Nausea

____

159. Alcohol consumption can cause ______________________ (hepatitis-B/cirrhosis).

160. Anabolic steroids help to maintain normal menstrual flow in sports females.

True ☐ False ☐

161. Drug abuse through ______________ route can cause serious infections like AIDS and hepatitis-B.

162. Drug and alcohol abuse may lead to stunted growth due to the premature closure of growth centres of long bones.

True ☐ False ☐

163. Among the listed factors, how many can help to prevent and control drug and alcohol abuse?

Disciplined parenting, Exerting peer pressure, Education, Looking for danger signs

____

164. Can psychologist and psychiatrists help to get rid of drugs and alcohol completely.

Yes ☐ No ☐

Answers

1. Health
2. 4 (Yoga, balanced diet vaccination and water).
3. *A* – Congenital disease,
   B – Infectious disease,
   C – Non-contagious,
   D – Cancer
4. Infectious
5. Pathogens
6. True
7. *Salmonella typhi*
8. (i) Small intestine, loss of appetite, intestinal perforation, Widal
   (ii) Mary Mallon
9. (i) *Streptococcus pneumoniae* and *Haemophilus influenzae*
   (ii) Alveoli of the lungs.
   (iii) Bluish lips and nails, fever, cough and headache.
   (iv) By inhaling droplets/aerosols released by infected person, sharing utensils and other articles with infected person.
10. *A* - Weight loss, blood in sputum, cough, high fever.
    B - DOTS (Directly Observed Treatment), Antituberculosis drugs - streptomycin, pyrazinamide, rifampicin, ethambutol, etc.
    C - *Vibrio cholerae*
    D - Contaminated water (waterborne infection)
    E - *Clostridium tetani*
    F - Lock jaw, painful muscular spasms, heart failure, etc.
    G - *Bordetella pertussis,*
    H - Triple vaccine, i.e. DPT, Erythromycin
11. *Mycobacterium leprae*
12. Plague
13. Leprosy
14. *Corynebacterium diphtheriae*
15. DPT
16. (i) Bacillus Calmette -Guerin
    (ii) Diphtheria-Pertussis-Tetanus
    (iii) National Leprosy Eradication Programme.
17. Syphilis
18. (i) Rhinoviruses
    (ii) Nasal congestion and discharge, soar throat and hoarseness.
19. Lung fibrosis is odd one out, as rhinoviruses infect nose and respiratory passage, but not the lungs.
20. Corona viruses
21. A - 4, B -1, C - 2, D - 3, E - 5
22. False. Salk vaccine is injectable whereas the sabin vaccine is given orally to prevent poliomyelitis.
23. False. *Varicella* vaccine helps to prevent chickenpox.
24. True
25. Chikungunya virus is odd one out, as it is transmitted through *Aedes* mosquito while the other three can be transmitted through semen of an infected male.
26. Malaria
27. *Plasmodium falciparum*
28. *A* - *P. vivax,*
    B - *P. ovale,*
    C - 9-14 day,
    D - 18-40 days,
    E - Malignant tertian or subtertian malaria
    F - Quartan malaria
29. Female *Anopheles* mosquito
30. (i) False. Only asexual life cycle of *Plasmodium* occurs in human whereas the sexual cycle occurs in *Anopheles* mosquito.
    (ii) True
31. Sporozoites
32. Sporozoites, asexually, merozoites
33. III $\rightarrow$ II $\rightarrow$ I $\rightarrow$ IV
34. Haemozoin (toxin)
35. Stomach
36. *A* - Salivary glands of mosquito,
    B - Sporozoites
37. True
38. *Entamoeba histolytica*
39. A - 2, B - 1, C - 4, D - 3
40. 2 (Giardiasis and amoebiasis)
41. (i) Blood sucking tse-tse fly of genus–*Glossina.*
    (ii) Sandflies of genus–*Phlebotomus*
    (iii) Trypomastigote form of reduviid bugs, *Panstrongylus.*
42. Intestinal
43. Muscular pain, anaemia, blockage of intestinal passage. (any two)
44. (i) Filariasis or elephantiasis
    (ii) *Wuchereria bancrofti and W. malayi*
    (iii) Bite by female *Culex* (vector) which carries immature larvae - microfilariae.
45. Taeniasis is caused by two species of *Taenia. T. solium* is found in pig whereas the cattle host *T. saginata.*
46. Cysticercus.
47. 3 (All of them) These fungi are collectively called dermatophytes.
48. Hot and moist.
49. *Trichophyton*
50. *Blastomyces,* it affects lungs, meninges, etc.
51. A - 3, B - 1, C - 2
52. (i) Eliminate malaria
    (ii) Smallpox
53. Immunity
54. Innate immunity
55. True
56. Mucus is odd one out, as it is coating of respiratory tract. Rest three form physiological barriers.
57. Polymorpho-nuclear leukocytes
58. Cellular
59. Interferons
60. Cytokine barrier
61. A - 4, B - 1, C - 2, D - 3
62. Acquired
63. False. An attack by a pathogen for the first time produces low intensity primary response.

64. Secondary response

65. (i) *A* - Heavy chain, *B* - Light chain, *C* - Antigen binding site
(ii) Antibody molecule
(iii) B - lymphocytes

66. H_2L_2

67. Variable, constant

68. *A* - 80%
B - First antibody to be produced in response of an infection.
C - IgA
D - IgE
E - Less than 1%

69. IgG

70. IgM

71. False. Antibodies formed by identical B-cells are called monoclonal antibodies.

72. Antigens or immunogens.

73. (i) Agglutination
(ii) Opsonisation
(iii) Precipitation

74. *A*–Humoral immune response
B–Cell-mediated immune response

75. Xenograft (graft from other species).

76. Cell-mediated

77. 4 (All of them).

78. Active immunity.

79. False. Active immunity can be generated in response to living or dead pathogen, proteins, etc.

80. Immunisation vaccine

81. Readymade antibodies

82. TAB vaccine is odd one out, as it provides active immunity whereas the rest two provide passive immunity.

83. IgA

84. Polyclonal, short.

85. Generate memory cells

86. True

87. Passive

88. Antigenic polypeptides of pathogen

89. Allergy

90. Pollen, mites, animal dander.

91. *A*- IgE antibody, *B*- Histamine and serotonin

92. True

93. Autoimmunity

94. SCID is odd one out, as it is an immunodeficiency disease while the rest are autoimmune diseases.

95. Primary lymphoid organs

96. (i) Primary, (ii) Thymus

97. 5 (All of them except thymus, it is primary lymphoid organ).

98. (i) *A* - Thymus, *B* - Lymph nodes, *C* - Lymphatic vessels.
(ii) *B*, as it is secondary lymphoid structure.
(iii) *A*, as it is primary lymphoid structure.

99. A - 2, B - 3, C - 4, D - 1

100. (i) Mucosal Associated Lymphoid Tissue (MALT), 50
(ii) Spleen, phagocytes.

101. False. AIDS is acquired during the lifetime of an individual.

102. Deficiency, syndrome, 1981.

103. Human Immunodeficiency Virus (HIV)

104. Retrovirus

105. Enveloped RNA

106. Sharing articles with infected person is odd one out, as it does not spread AIDS, while the other listed modes can spread/transmit AIDS through body fluids.

107. (i) *A*–Retrovirus , *B* –Viral DNA is produced by reverse transcriptase.
(ii) Reverse transcriptase
(iii) Macrophages
(iv) *C*–Host DNA, *D*–Viral RNA is produced by infected cell.
(v) False. Infected cell can survive even after the replication and release of virus from them.

108. Macrophages

109. Helper T-lymphocytes.

110. Helper T-lymphocytes decreases.

111. Enzyme Linked Immuno-Sorbent Assay (ELISA).

112. (i) National AIDS Control Organisation.
(ii) Non-Governmental Organisation.

113. 3 (All of them except isolation of AIDS infected person, it is not a preventive measure for AIDS).

114. Cancer

115. Contact inhibition, tumours

116. *A*-Benign tumour, *B*-Sarcoma.

117. Malignant tumour

118. A-3, B-1, C-4, D-2

119. UV rays is odd one out, as these are non-ionising carcinogenic radiations, while the other two are ionising carcinogenic radiations.

120. Chemical

121. Oncogenic viruses

122. Biological

123. Protooncogenes or cellular oncogenes.

124. Leukemia

125. A-3, B-1, C-2, D-4

126. True

127. Radiotherapy

128. Alpha (α) - interferons

129. Chemotherapy

130. Drugs

131. Opioids, cannabinoids, coca alkaloids.

132. False. Majority of drugs are obtained from flowering plants and some from fungi.

133. Gastrointestinal tract

134. Heroin

135. Diacetylmorphine

136. (ii) represents the correct structure of morphine.

137. (i) Opium poppy or *Papaver somniferum.*
(ii) Morphine
(iii) Latex of the plant

138. Orally is odd one out, as heroin is not taken orally.

139. True

140. Pethidine/methadone (any one).

141. Brain

142. Cannabinoids are obtained from inflorescence, flower tops, leaves and resin of the plant *Cannabis sativa.*

143. Cannabinoids

144. *A* - Hydroxyl (OH) group

145. *Cannabis sativa,* cardiovascular

146. *Erythroxylum coca*

147. A-3, B-1, C-2

148. Dopamine, stimulates

149. (i) *Datura*
(ii) Hallucinogenic
(iii) *Atropa belladonna* and *Erythroxylum* (in excess).

150. 3 (All of them except hashish, it is not used as medicinal drug).

151. Cannabinoids

152. Lysergic Acid Diethylamides

153. (i) Nicotine
(ii) Adrenaline, nor-adrenaline
(iii) Carbon monoxide (CO)

154. Excitement

155. Mental illness

156. Addiction

157. Dependency

158. Euphoria is odd one out, as it is a sense of happiness and is associated with drug/alcohol consumption.

159. Cirrhosis

160. False. Anabolic steroids are misused by sports persons. In females, these drugs cause abnormal menstrual cycle, facial hair growth, etc.

161. Intravenous

162. True

163. 3 (All of them except exerting peer pressure). One should avoid undue peer pressure to refrain from drug and alcohol abuse.

164. Yes

CHAPTER NINE

STRATEGIES FOR ENHANCEMENT IN FOOD PRODUCTION

Topic 1 : Animal Husbandry, Management of Farms and Farm Animals

1. The new techniques like ____________ and ____________ are used to enhance food production.

2. Tissue culture deals with the sterile isolation and *in vitro* growth of an embryo with the ultimate objective of obtaining a viable plant and increasing food production.

   True ☐ False ☐

3. Give a biological term for following statements.
   (i) Animals domesticated by humans for commercial purposes.____________
   (ii) Name the agricultural practice of breeding and raising livestock. ____________

4. More than 30 per cent of the world's livestock population is from India and China.

   True ☐ False ☐

5. The contribution of India and China to the world farm produce is only 25 per cent despite having a large proportion of livestock population. (*Give reason*)

   ____

6. For improving quality and productivity of animal breeding conventional methods should be continued.

   True ☐ False ☐

7. Give a biological term for the following sentence.

   The management of animals for milk and its products.

   ____

8. Increased yield and improved quality of milk are main objectives of animal husbandry.
(*Rewrite the statement by changing the underlined word.*)

____

9. Milk yield is dependent on ______________________________ .

10. Cattle animals with high yield potential and disease sensitivity are selected for cattle farming.

True ☐ False ☐

11. Stringent cleanliness and hygiene of only the cattle is essential in dairy farm management.
(*Rewrite the statement by changing the underlined word.*)

____

12. What are essential practices in dairy farm management?
(Veterinary checks / Conventional breeding)

____

13. Milch breeds are dual purpose breeds where females are good milk producers and bullocks are good labourers.

True ☐ False ☐

14. Pick the odd one out from the breeds of cattle.

Malvi, Red Sindhi, Nageri, Kangayam

____

15. Match the following columns.

Column I (Breeds)	Column II (Animals)
A. Gaddi	1. Pigs
B. Berkshire	2. Cows
C. Lohi	3. Goats
D. Holstein-Friesian	4. Sheep

A.__________ B.__________ C. __________ D. __________

16. How many of the following are Indian breeds of buffaloes?

Kankrej, Deoni, Jaffrabadi, Manipuri, Surti, Mehsana, Nellore, Beetal, Bhadawari

____

17. Match the following columns.

Column I (Breed names)	Column II (Characteristics)
A. Gir	1. Drought breed
B. Tharparkar	2. Milch breed
C. Nageri	3. General utility

A.__________ B.__________ C. __________

18. A bacterial disease in cattle animals is _______________ (anthrax/coccidiosis).

19. Match the following columns.

Column I (Diseases)	Column II (Causal organisms)
A. Mastitis	1. Rinderpest virus
B. Tick fever	2. *Fasciola hepatica*
C. Cattle plague	3. *Corynebacterium pyogenes*
D. Liver rot	4. *Babesia bigemina*

A. _______________ B. _______________ C. _______________ D. _______________

20. The class of domesticated birds used for food or eggs is _______________.

21. How many of these animals are include in poultry?

Chicken, Ducks, Turkey, Sparrow, Bee and Geese

____

22. Complete the following analogy.
Egg laying birds : _______________ : : Birds for meat : Broilers

23. Pick the odd one out from the following breeds.

Aseel, White leghorn, Malay and Ghagus

____

24. Chittagong is a variety of <u>fish</u>.
(*Rewrite the statement by changing the underlined word.*)

____

25. How many of the following are exotic breeds of poultry?

Busra, Rhode Island Red, Barred Plymouth Rock, Australorp, New Hampshire, Dahlem Red, Kadaknath.

____

26. _______________ (ILS 82/ Langshan) is a crossbreed variety of poultry bird developed in India.

27. The fowls are reared to obtain how many of the products mentioned below?

Eggs, Shagreen, Meat, Feathers, Manure and Milk

____

28. The 'bird flu infection' which created a scare in the country, drastically affected the _______________ consumption and is known to be caused by a _______________.

29. Match the following columns.

	Column I (Diseases of poultry)		Column II (Pathogens)
A.	Mycosis	1.	*Aspergillus*
B.	Aspergillosis	2.	*Pasteurella*
C.	Infectious coryza	3.	*Candida albicans*
D.	Pullorum disease	4.	*Haemophilus gallinarum*
E.	Cholera	5.	*Salmonella*

A.__________ B.__________ C. __________ D. __________
E.__________

30. An infected bird keeps their beaks open, have fever with yellowish white diarrhoea. The disease is ____________ (Newcastle disease/ Marck's disease).

Topic 2 : Animal Breeding, Bee-Keeping and Fisheries

31. Give scientific term for the following statements.

(i) The practice of science which deals with the selection of animals having superior characters in terms of growth rate and productivity. ____________

(ii) A group of animals related by descent and similar in most characters. ____________

32. The best known improved breed of cow is ____________

33. Two animals of the same breed will share how many of the following characteristics?

General appearance, Features, Size, Configuration, Weight, Hair colour

____

34. What is common to the given processes? (Inbreeding, Outbreeding, Artificial insemination). ____________

35. Complete the following process of inbreeding.

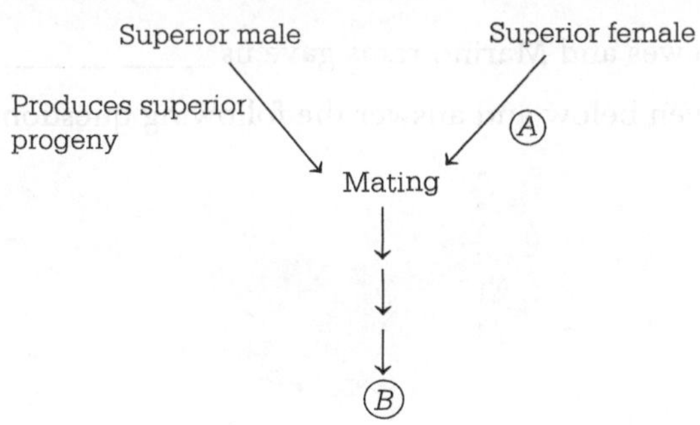

A.__________ B. __________

36. How does inbreeding differ from outbreeding methods of animal breeding? (*Answer in one line.*)

____

37. Name the condition, in which continued inbreeding reduces fertility of an animal.

____

38. If we want to evolve a pureline in any animal, which technique of breeding should we adopt?

____

39. What would be the impact of continuous inbreeding in a population containing harmful genes?

____

40. How many of the following methods can be used to overcome inbreeding depression.

Outcrossing, Incomplete dominance, Cross breeding and Interspecific hybridisation, MOET

____

41. Match the following columns.

Column I (Breeding methods)	Column II (Processes)
A. Cross breeding	1. Male and female animals of two different species are mated
B. Interspecific hybridisation	2. Breeding between different breeds
C. Outcrossing	3. Mating of animals within the same breed, but having no common ancestors for 4-6 generations

A.______________ B.______________ C. ______________

42. The best breeding method for animals that are below average in productivity and growth rate is interspecific hybridisation.
(*Rewrite the statement by changing the underlined word.*)

____

43. Through which process the desirable qualities of two different breeds can be combined?

____

44. Crossing Bikaneri ewes and Marino rams gave us ______________ .

45. Study the figure given below and answer the following questions.

(i) The figure above represents the progeny of ______________________ and ______________________ called ______________________.

(ii) This type of breed is obtained using cross breeding.

True ☐ False ☐

(iii) Mules are <u>sexually active</u> hybrids.
(*Rewrite the statement by changing the underlined word.*)

____

46. Artificial breeding of cattle is brought about by ____________ (artificial insemination/ homozygotic twinning).

47. In cross breeding, semen is collected from selected male is directly injected into selected female.

True ☐ False ☐

48. The first artificial insemination was performed at ________________ (IVRI/ NDRI).

49. The semen from superior male can be used to fertilise a single female.

True ☐ False ☐

50. Pick the odd one out.

Inbreeding, MOET, Outcrossing, Intraspecific hybridisation

____

51. The following chart depicts the methodology which is employed in cattle improvement.

Selected animal is administered with hormone (X)

↓

(*A*)

↓

6 - 8 eggs per cycle

↓

(*B*)

↓

Recovery of embryo at 8-32 cell stage.

↓

(*C*)

(i) Which hormone is used in step 1 and why? ______________________

(ii) These steps are part of which technique ? (Transgenesis/ MOET) ____________

(iii) Identify the steps labelled *A, B* and *C.*

A. ____________ B. ____________ C. ____________

52. Expand MOET. ______________________________________

53. MOET is carried out mainly for ______________________________

54. MOET has been performed on how many of the following animals?

Cattle, Dogs, Sheep, Rabbits, Buffaloes, Mares, Cats

____

55. Give one word for the following statement.

The maintenance of hives of honeybees for the production of honey.

____

56. Match the following columns.

Column I (Species of honeybees)	Column II (Characteristics of species)
A. *Apis dorsata*	1. Rarely sting
B. *Apis florea*	2. Easily domesticated
C. *Apis mellifera*	3. Giant bee
D. *Apis indica*	4. European bee

A.__________ B.__________ C. __________ D. __________

57. What is common to all of these products?

Honey, Venom and Beeswax, Propolis

____

58. Complete the following analogy.
Apis florea : Little bee :: __________ : Rock bee.

59. Bee-keeping can be practiced in any area where there are sufficient. Bee pastures of some __________ (fruit orchards and cultivated crops / rotten wood).

60. The most common species for apiculture is *Apis dorsata*.

True ☐ False ☐

61. Beehives are kept in crop fields during flowering period. (*Give reason*)

____

62. Bees are known pollinators of __________ (*Brassica* /rose).

63. How many of the following components of social organisation include in a beehive ?

Queen, Drone, King, Worker and Servants

____

64. The drones develop from haploid eggs, which are not fertilised. This development is termed as __________ .

65. For bee-keeping selection of suitable location is not a primary concern.

True ☐ False ☐

66. Complete the following processes which are important for successful bee-keeping.

Knowledge of the (A) and habits of bee
↓
Selection of location
↓
Catching and hiving of (B)
↓
Handling and collection of (C) and (D)

A.______________ B.______________ C. ______________ D. ______________

67. Complete the following analogy with respect to bee diseases.
Acarine : ____________________ : : Nosema : *Nosema apis*

68. Give scientific term for the following statement.

The industry of catching, processing or selling of fish or other aquatic animals.

____

69. How many of the following are associated with the fisheries industry?

Fish, Prawn, Crab, Lobster, *Hydrilla*, Edible oyster, Molluscs, Corals

____

70. Pick the odd one out.

Catla, *Hilsa*, Rohu, Carp

____

71. Match the following columns.

Column I (Names of breed)	Column II (Scientific names)
A. Mrigal	1. *Heteropneustes fossilis*
B. Singhi	2. *Cirrhinus mrigala*
C. Magur	3. *Mystus sengala*
D. Singhara	4. *Clarias batrachus*
E. Sardine	5. *Stromateus sinensis*

A.______________ B.______________ C. ______________ D. ______________
E.______________

72. What is common to following fishes?

Eel, Sardines, Mackerel and Pomfrets

____

73. Give a biological term for the following statements.

(i) The scientific rearing and management of fishes under controlled conditions. ____________

(ii) The farming of freshwater and saltwater organisms. ____________

(iii) Fishes raised in ponds and tanks. ____________

(iv) Management of catching fishes without raising them. ____________

74. Complete the following analogy with respect to revolutions occurred in different field.
Green revolution : Agriculture :: Blue revolution : ____________

75. Certain carnivorous fishes ____________ (*Gambusia/Carasinus*) help in disease control.

76. Complete the following analogy.
Fish oil : ____________ :: Fish glue : Skin of cod.

77. A substance obtained from swim bladders of fish used in clarification of vinegar, wine, etc., is ____________ (shagreen/isinglass).

78. Match the following columns.

	Column I (Processes for rearing animals)		Column II (Scientific name of organisms)
A.	Sericulture	1.	*Tachardia lacca*
B.	Lac culture	2.	*Oystreidae*
C.	Mariculture	3.	*Bombyx mori*

A.____________ B.____________ C. ____________

Topic 3 : Plant Breeding and Single Cell Protein

79. ____________ was responsible for our country to meet the national requirements in food production and also helped us to export it.

80. Give one word for the following statement.
The purposeful manipulation of plant species towards a desired outcome.

____

81. Plant breeding aims towards ____________ along with ____________.

82. Steps of classical plant breeding is hybridisation of purelines followed by ____________.

83. Select the traits that breeders have tried to incorporate into crop plants from those given below.

Increased crop yield, Improved quality, Increased tolerance to environmental stresses, Foliage, Resistance to pathogens, Increased tolerance to insect pests, Height

____

84. Arrange the following steps of plant breeding programmes in sequence.
I. Cross hybridisation between selected parents.
II. Collection of variability.
III. Testing, release and commercialisation of new cultivars.
IV. Evaluation and selection of parents.
V. Selection and testing superior recombinants.

____

85. Give scientific term for the following statement.
The entire collection of plants having all the diverse alleles for all genes in a given crop.

____

86. In most plants, the pre-existing genetic variability is derived from _______________ of the crop.

87. Purelines always possess _______________ (heterozygosity/ homozygosity).

88. What is the need for evaluation and selection of parents?

____

89. High protein quality of one parent is needed to be combined with disease resistance from another parent. How will we achieve this?

____

90. Match the following columns.

Column I (Types of crosses)	Column II (Processes carried out)
A. Intervarietal	1. Cross between two varieties of same crop
B. Intravarietal	2. Cross between different genotypes of the same variety
C. Interspecific	3. Cross between two different species of a genus
D. Intergeneric	4. Cross between two genera

A. _______________ B. _______________ C. _______________ D. _______________

91. During hybridisation, pollen grains from selected male plant are collected and placed in the style of flowers of female plant.
(*Rewrite the statement by changing the underlined word.*)

____

92. _______________ is a process where anthers are physically removed in self-pollinated plants.

93. What is the process called where flower buds are artificially enclosed to avoid undesired pollination? _______________

94. In cross hybridisation, all hybrids have combined the desirable characters of two parents.

True ☐ False ☐

95. The selection process yields plants that are ______________________ to both of the parents.

96. The progeny obtained in the selection is self-pollinated for several generations to attain the state of homozygosity. (*Give reason*)

____

97. How many of the following traits are evaluated for testing of new cultivars?

Yield, Quality, Disease resistance, Pollination efficiency

____

98. New cultivars are tested for five growing seasons at several locations in the country. (*Rewrite the statement by changing the underlined word.*)

____

99. Agriculture accounts for approximately 23 per cent of India's GDP.

True ☐ False ☐

100. During the period 1960 to 2000, wheat production increased from 11 million tonnes to 75 million tonnes, while rice production went up from 35 million tonnes to 89.5 million tonnes in India. (*Give reason*)

____

101. *Triticale* is developed by a cross of the plants ______________________.

102. Semidwarf varieties were developed by______________________ (Norman E Borlaug/ MS Swaminathan).

103. What is common to Sonalika and Kalyan Sona?

____

104. IR-8 was developed at the International Centre for Wheat and Maize Improvement.

True ☐ False ☐

105. Complete the following analogy with respect to origin of variety.

______________ : Taiwan :: Jaya : India

106. Complete the following analogy with respect to origin of variety.

Saccharum barberi : North India :: ______________: South India

107. *Saccharum officinarum* had poor sugar content and yield.
(*Rewrite the statement by changing the underlined word.*)

____

108. *Saccharum barberi* and *Saccharum officinarum* when crossed with each other would yield a progeny with better traits. Identify these traits.

____

109. Indian hybrid crop of high yielding varieties have been developed for how many of the following crops?

Maize, Rice, Sugarcane, Wheat, *Parthenium*, Millets

____

110. Crop losses can often be significant, i.e. up to 20-30 per cent or sometimes even total. How can this loss be overcome to enhance food production?

____

111. Resistance of the host plant is dependent on the ______________ of the host plant.

112. Complete the following table given below.

Crops	Varieties	Resistance to diseases
Chilli	(A)	Chilli mosaic virus
(B)	Pusa Swarnim	(C)
Wheat	(D)	Leaf and stipe rust
Cowpea	Pusa Komal	(E)
(F)	Pusa Shubhra	(G)

A. __________ B. __________ C. __________ D. __________

E. __________ F. __________ G. __________

113. The traditional breeding for disease resistance is performed using mutations.

True ☐ False ☐

114. Only a limited number of disease resistance genes are present and identified in a crop plant.

True ☐ False ☐

115. Give a biological term for the following statement.
The process by which genetic variations are created.

____

116. The changes in the base sequence within genes results in __________ .

117. Use of induce mutation in plant breeding is to develop plants that have desirable characters is called ______________ .

118. Pick the odd one out.

Nitrous acid, Maleic hydrazide, Gamma radiation, Methyl methane sulphonate

____

119. In mung bean, resistance to chilli mosaic virus and leaf curl were induced by mutations. (*Rewrite the statement by changing the underlined words.*)

____

120. Resistance to tobacco mosaic virus was developed in *Abelmoschus esculentus*. (*Rewrite the statement by changing the underlined word.*)

____

121. Parbhani Kranti is a disease resistant variety of *Brassica*.

True ☐ False ☐

122. Transfer of genes for resistance is achieved by________________ between the target and the source plant followed by selection.

123. Sources of resistance genes may be ____________________ (germplasm collections/conventional breeds) of the crop or wild relatives.

124. Complete the following table given below.

Crops	Characteristic feature which imparts resistance to insect pests	Insect pests
Cotton	(A)	(B)
(C)	Solid stems	Stem sawfly

A.____________ B.____________ C. ____________

125. In maize, resistance to stem borer is attributed to low aspartic acid, high nitrogen and sugar.

True ☐ False ☐

126. Thorny leaves in a several plants are associated with resistance to insect pests. (*Rewrite the statement by changing the underlined word.*)

____

127. Complete the following table given below.

Crops	Hybrid insect pests resistant varieties	Insect pests
(A)	Pusa Gaurav	(B)
Flat bean	Pusa Sem 2	(C)
Okra	(D)	Shoot and fruit borer

A.____________ B.____________ C. ____________ D. ____________

128. Plant breeding for improved food quality is called ________________ (micropropagation / biofortification).

129. ________________ occurs when there is the absence of essential nutrients in the diet.

130. Breeding for improved nutritional quality is undertaken with the objectives of improving the content of.

Protein, Oil, Vitamin, Minerals, Cholesterol, Essential micronutrients, Hormones

____

131. Amino acid, lysine and tryptophan rich crop variety is ____________ (maize/millets).

132. ________________ is a high protein rich variety of wheat.

133. Give two examples of each of the following.

(i) Vitamin-A enriched crops ______________________

(ii) Vitamin-C enriched crops ______________________

(iii) Calcium and iron enriched corps ______________________

(iv) Protein enriched crops ______________________

134. Name an alternate source of proteins for animal and human nutrition.

____

135. Why is a microorganism like *Methylophilus methylotrophus* grown on an industrial scale as a source of good protein?

____

136. Complete the following table given below.

Microorganisms (Source of protein)	Substrates	Classes
Methylophilus methylotrophus	(A)	Bacteria
(B)	Straw, molasses, animal manure	Algae
Agaricus	(C)	(D)
(E)	Molasses	Yeast

A.__________ B.__________ C. __________ D.__________

E. __________

Topic 4 : Tissue Culture

137. Give a biological term for.

The growth of cells in an artificial medium.

____

138. Any part of a plant that can regenerate into a new plant in a test tube under sterile condition in special nutrient media is called __________. (tissue/ explant)

139. The nutrient medium prepared for tissue culture should have how many of these components?

Sucrose, Inorganic salts, Vitamins, Amino acids, Growth regulators, Agar

____

140. Name the biological term for the ability of single cells to divide and produce all the differentiated cells in the organism. _______________

141. The technique of tissue culture was started by PK Maheshwari in 1903.
(*Rewrite the statement by changing the underlined word.*)

____

142. Callus is an _______________ (undifferentiated/differentiated) mass of cells obtained in tissue culture.

143. Given below are steps involved in the process of tissue culture in plants, arrange them in the correct sequence of events.

I. Selection of plant, II. Incubation, III. Hardening, IV. Isolation of explant. V. Sub-culturing, VI. Initiation of callus, VII. Sterilisation of explant, VIII. Inoculation of explant, IX. Regeneration.

____

144. Define cybrid. _______________

145. Match the following columns.

	Column I (Plant tissue culture techniques)		Column II (Applications)
A.	Micropropagation	1.	Artificial seed production
B.	Somatic embryogenesis	2.	Haploid plant production
C.	Somaclonal variation	3.	Production of disease resistant plants
D.	Meristem culture	4.	Rapid propagation of identical plants
E.	Anther culture	5.	Create variations in plants

A. _______________ B. _______________ C. _______________ D. _______________
E. _______________

146. _______________ plants will be genetically identical to the original plants from which they were grow.

147. Name any three food plants which are produced on commercial scale by micropropagation.

____

148. For the recovery of healthy plants from diseased plants, the explant should be derived from the stem of a plant
(*Rewrite the statement by changing the underlined word.*)

____

149. Consider the figure and answer the questions that follows.

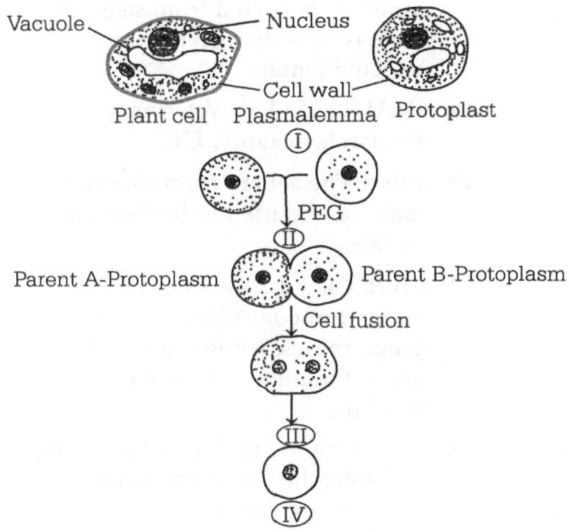

(i) Mention the combination of enzymes used at step I.

____

(ii) At step II, which chemical mediates the fusion of protoplasts A and B?

____

(iii) Cell fusion has already occurred, then what does stage III indicate?

____

(iv) The cell formed at stage IV is ________. This process is known as ________.

150. What happens when a protoplast of tomato is fused with the protoplast of potato and cultivated?

____

Answers

1. Animal husbandry, plant breeding.
2. False. Embryo culture deals with the sterile isolation and *in vitro* growth of an embryo with the ultimate objective of obtaining a viable plant and increasing food production.
3. (i) Livestock
   (ii) Animal husbandry
4. False. More than 70 per cent of the world livestock population is from India and China.
5. The reason for lower contribution of these countries is low productivity due to the use of inferior breeds and inadequate food.
6. False. For improving quality and productivity of animal breeding modern methods should be continued.
7. Dairying
8. Dairy farm management.
9. The quality of breeds in the farm.
10. False. Cattle animals with high yield potential and disease resistance are selected for cattle farming.
11. The cattle and the handlers.
12. An essential practices in dairy farm management is veterinary checks.
13. False. General utility breeds are dual purpose breeds where females are good milk producers and bullocks are good labourers.
14. Red Sindhi is odd one out breed as it is a milch breed, while others are drought breeds.
15. A-3, B-1, C-4, D-2
16. 4 (Jaffrabadi, Surti, Mehsana, Bhadawari).
17. A-2, B-3, C-1
18. Anthrax
19. A-3, B-4, C-1, D-2
20. Poultry
21. 4(Chicken, ducks, turkey and geese).
22. Layers
23. White leghorn is odd one out as it is an exotic breed of poultry, while others are indigenous breeds.

24. Poultry bird
25. 5 (Rhode Island Red, Barred Plymouth Rock, Australorp, New Hampshire, Dahlem Red).
26. ILS 82
27. 4 (Eggs, meat, feathers and manure).
28. Egg and chicken, H5N1 virus.
29. A-3, B-1, C-4, D-5, E-2
30. Newcastle disease
31. (i) Animal breeding
(ii) Breeds
32. Jersey
33. 4 (General appearance, features, size, configuration).
34. All of the given processes are the methods of animal breeding.
35. *A*–Produces more milk per lactation
B–Progeny
36. In inbreeding, more closely related individuals within the same breed are mated for 4-6 generations, while in out breeding unrelated animals of same or different breeds are mated.
37. Inbreeding depression
38. Inbreeding because it increases homozygosity.
39. Due to inbreeding, the deleterious (harmful) recessive genes will accumulate leading to homozygous recessive (i.e. lethal) condition.
40. 3 (Outcrossing, cross breeding and interspecific hybridisation).
41. A-2, B-1, C-3.
42. Outcrossing
43. Cross breeding
44. Hisardale
45. (i) Male donkey, female horse, mule.
(ii) False. This type of breed is obtained using interspecific hybridisation.
(iii) Sterile
46. Artificial insemination
47. False. In artificial insemination, semen is collected from selected male is directly injected into selected female.
48. IVRI, i.e. Indian Veterinary Research Institute, UP.
49. False. The semen from superior male can be used to fertilise many females.
50. MOET is odd one out as it is a method of controlled breeding programmes, while others are normal methods of animal breeding.
51. (i) Hormone is FSH for inducing follicular maturation and superovulation.
(ii) These steps are involved in MOET of controlled animal breeding programmes.
(iii) *A*–Superovulation due to induced follicular maturation, *B*–Artificial insemination, *C*–Transfer of embryo to surrogate.
52. Multiple Ovulation Embryo Transfer technology.
53. Herd improvement
54. 5 (Cattle, sheep, rabbits, buffaloes, mares).
55. Apiculture
56. A-3, B-1, C-4, D-2
57. All of these products are derived from honeybees in apiculture.
58. *Apis dorsata*
59. Fruit orchards and cultivated crops.
60. False. The most common species of honeybee for apiculture is *Apis indica*.
61. Keeping beehives in crop fields during flowering period to increase pollination efficiency and improves the yield of both the crops and honey.
62. Bees are known pollinators of *Brassica*, along with apples, pears and sunflower.
63. 3 (Queen, drone and worker).
64. Parthenogenesis
65. False. Selection of a suitable location for keeping the beehives is very essential for apiculture.
66. *A*–Nature, *B*–Swarms, *C*–Honey, *D*–Wax
67. *Acarapis woodi*
68. Fisheries
69. 6 (Fish, prawn, crab, lobster, edible oysters and molluscs).
70. *Hilsa* is the odd one out as it is a marine water fish, while others are freshwater fishes.
71. A-2, B-1, C-4, D-3, E-5
72. They all are marine water fishes.
73. (i) Pisciculture
(ii) Aquaculture
(iii) Culture fishery
(iv) Capture fishery
74. Aquaculture
75. *Gambusia*
76. Shark liver
77. Isinglass
78. A-3, B-1, C-2
79. Green revolution
80. Plant breeding
81. High yield, disease resistance.
82. Artificial selection
83. All characters except foliage and height.
84. The correct sequence of plant breeding programme is II→IV→I→V→III.
85. Germplasm
86. Wild varieties
87. Homozygosity
88. To identify plants with desirable combination of characters.
89. By cross hybridising, the two parents to produce hybrids that genetically combine the desired characters.
90. A-2, B-3, C-1, D-4

91. On the stigma

92. Emasculation

93. Bagging

94. False. The hybrids do not necessarily combine the desirable characters of selected parents.

95. Superior

96. This is done so that characters do not segregate in the progeny.

97. 3 (Yield, quality and disease resistance).

98. At least three.

99. False. Agriculture accounts for approximately 33 per cent of India's GDP.

100. This was due to the development of semidwarf varieties of wheat and rice.

101. *Triticum aestivum* and European rye (*Secale cereale*).

102. Norman E Borlaug

103. They were high yielding and disease resistant semidwarf varieties of wheat.

104. False. IR-8 was developed at the International Rice Research Institute (IRRI), Philippines.

105. Taichung Native -1

106. *Saccharum officinarum*

107. *Saccharum barberi*

108. The desirable qualities were high yield, thick stems, high sugar content and ability to grow in North India.

109. 5 (Maize, rice, wheat, millets and sugarcane). *Parthenium* is a weed.

110. In such a situation, breeding and development of cultivars resistant to disease helps to enhance the food production.

111. Genetic constitution

112. *A*–Pusa Sadabahar, *B*–*Brassica*, *C*–White rust, *D*–Himgiri, *E*–Bacterial blight, *F*–Cauliflower, *G*–Black rot/Curl blight black rot.

113. False. The traditional breeding for disease resistance is performed using hybridisation and selection.

114. True

115. Mutation

116. New traits.

117. Mutation breeding

118. The gamma radiation is odd one out as it is a physical mutagen, while others are chemical mutagens.

119. Yellow mosaic virus and powdery mildew

120. Yellow mosaic virus

121. False. Parbhani Kranti is a disease resistant variety of bhindi (*A. esculentus*).

122. Hybridisation

123. Germplasm collections

124. *A*–Hairy leaves, *B*–Jassids and *C*–Wheat.

125. False. In maize, resistance to stem borer is attributed to high aspartic acid, low nitrogen and sugar content.

126. Hairy leaves

127. *A*–*Brassica* , *B*–Aphids, *C*–Jassids, aphids and fruit borer, *D*–Pusa Sawani, Pusa A–4.

128. Biofortification

129. Hidden hunger

130. 5 (Protein, oil, vitamin, minerals and essential micronutrients).

131. Maize

132. Atlas 66

133. (i) Carrots, pumpkin
(ii) Bitter gourd, mustard
(iii) Spinach, bathua
(iv) Lablab, garden peas

134. Single Cell Protein (SCP)

135. Because of its high rate of biomass production and growth. This bacterium can produce approximately 25 tonnes of protein.

136. *A*–Methanol, *B*–*Spirulina* , *C*–Compost, *D*–Fungi, *E*–*Saccharomyces cerevisiae.*

137. Tissue culture

138. Explant

139. 6 (All of them)

140. Totipotency

141. Haberlandt in 1902.

142. Undifferentiated.

143. The correct sequence is I → IV → VII → VIII → II → VI → V → IX → III.

144. A cybrid is a hybrid carrying cytoplasm of two different plants and genome of one plant.

145. A-4, B-1, C-5, D-3, E-2

146. Somaclonal

147. Tomato, banana and apple.

148. Meristem

149. (i) Cellulase and pectinase
(ii) PEG, i.e. Poly Ethylene Glycol
(iii) At stage III, nuclear fusion is taking place.
(iv) Somatic hybrid, somatic hybridisation.

150. It results in the production of a hybrid called pomato. This plant did not have all the desired characters.

CHAPTER TEN

MICROBES IN HUMAN WELFARE

Topic 1 : Basic Concepts and Microbes in Household Products

1. The organisms like _______________________ are found on places like thermal vents where no other life form is found.

2. Identify the organisms shown in *A* and *B*.

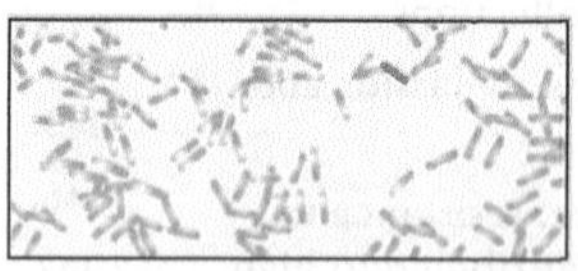

A

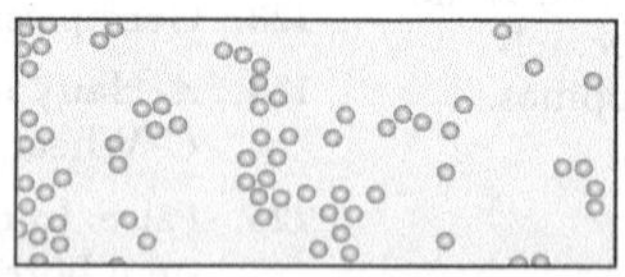

B

A.________________ B.________________

3. Some microbes, i.e. viruses cannot be grown on culture media are called ____________ (parasitic/obligate) organisms.

4. Choose the odd one out with respect to the type of microbes shown in the diagrams given below.

A

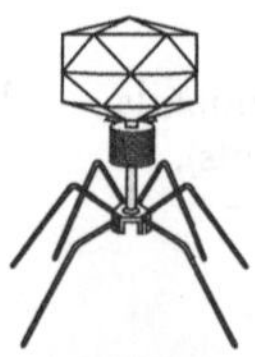

B

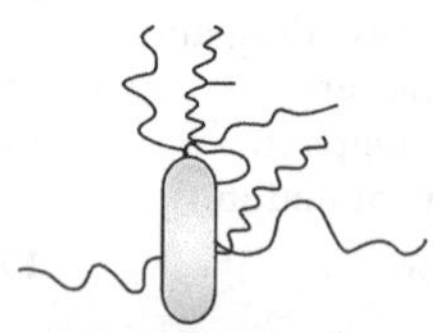

C

D

5. The microbes grown on nutritive media to form colonies cannot be seen with the naked eyes.

   True ☐ False ☐

6. The ________________ (bacteria/virus) and ______________ (fungi/lichen) are the organisms can be easily grown on culture media.

7. An example of rod-shaped virus is __________________________ (adenovirus / tobacco mosaic virus).

8. Name one virus that causes respiratory problem. ______________________________

9. Expand LAB. ____

10. Complete the flowchart depicting the formation of curd.

    LAB in inoculum ⟶ Produce an acid from milk sugar ____________ ⟶ Coagulate and partially digest milk protein ____________ ⟶ Curd

11. LAB improves the nutritional quality of curd by increasing vitamin-A. (*Rewrite the statement by changing the underlined word.*)

    ____

12. Match the following columns.

Column I (Household products)	Column II (Microbes)
A. Bread	1. *Propionibacterium shermanii*
B. Swiss cheese	2. *Streptococcus lactis*
C. Yoghurt	3. *Saccharomyces cerevisiae*
D. Butter	4. *Lactobacillus bulgaricus*

A. ________________ B. ________________

C. ________________ D. ________________

13. The puffing up of idli dough is caused due to the production of ________________ during ________________ .

14. Name the drink made by fermenting sap from palms in Southern India. ____________

15. The particular flavour of Roquefort cheese is imparted by a bacteria.

    True ☐ False ☐

Topic 2 : Microbes in Industrial Products

16. Beverages and antibiotics are produced on industrial scale in very large vessels called ____________________ .

17. A fermentation plant in shown in the figure given below. Name any two drinks which are formed in these fermentors.

____

18. Complete the following analogy given below with respect to bread making.

Bread : Baker's yeast : : ____________________ : Brewer's yeast

19. Complete the flowchart depicting the metabolic reactions involved in the production of ethanol by yeast.

$$\underset{\text{Sucrose}}{C_{12}H_{22}O_{11}} + Ⓐ \xrightarrow[\text{Invertase}]{\text{Yeast}} \underset{\text{Glucose}}{C_6H_{12}O_6} + Ⓑ$$

$$\downarrow \text{Yeast } Ⓒ$$

$$\underset{\text{(Ethanol)}}{2C_2H_5OH} + 2Ⓓ$$

A.________________ B.________________

C.________________ D.________________

20. Choose the odd one out from the words given below.

Whisky, Brandy, Rum, Wine

____

21. Complete the following table given below.

Alcoholic beverages	Raw materials	Alcoholic contents
Wine	Grapes	10-20%
Beer	(A)	4-8%
Rum	(B)	40%
(C)	Distillation of wine	43-57%

A. ______________ B. ______________ C. ______________

22. Give the biological term for the chemical substances, which are produced by some microbes and can kill or retard the growth of other microbes. ______________

23. The first antibiotic ______________ was discovered by Alexander Fleming, while working on ______________ bacteria.

24. Name the scientists who were awarded the Nobel Prize in 1945 for the discovery of antibiotic produced by the mould *Penicillium notatum*. ______________

25. Match the following columns.

Column I (Antibiotics)	Column II (Microbial sources)
A. Erythromycin	1. *Streptomyces rimosus*
B. Terramycin	2. *Streptomyces fradiae*
C. Streptomycin	3. *Bacillus licheniformis*
D. Bacitracin	4. *Streptomyces erythreus*
E. Neomycin	5. *Streptomyces griseus*

A. ______________ B. ______________

C. ______________ D. ______________

E. ______________

26. How many broad spectrum antibiotics are listed below?

Tetracycline, Kanamycin, Penicillin, Streptomycin, Chloramphenicol, Rifampicin, Ampicillin

____

27. Given below are name of certain diseases, name the antibiotics which is used to treat these disease.

(i) Plague ______________

(ii) Whooping cough ______________

(iii) Diphtheria ______________

(iv) Leprosy ______________

28. Citric acid is an organic acid which is produced by a bacterium.
(*Rewrite the statement by changing the underlined word.*)

____

29. Match the following columns.

	Column I (Organic acids)		Column II (Microbial sources)
A.	Acetic acid	1.	*Clostridium butylicum*
B.	Lactic acid	2.	*Acetobacter aceti*
C.	Butyric acid	3.	*Aspergillus niger*
D.	Citric acid	4.	*Lactobacillus*

A. ______________ B. ______________

C. ______________ D. ______________

30. Complete the following table given below.

Enzymes	Microbial sources	Functions
Lipase	(A)	Oil stain removal from laundry
Protease	*Bacillus subtilis*	(B)
(C)	*Streptococcus*	Clot buster
Glucose oxidase	(D)	To check blood glucose level
Pectinase	(E)	Retting of fibres

A. ______________ B. ______________ C. ______________

D. ______________ E. ______________

31. Read the following statements and identify the bioactive molecule.
I. Produced by fungus *Trichoderma polysporum.*
II. Used as an immunosuppressive agent in organ transplant patients.

____

32. Name the microbial sources of statins. ______________

33. Statins are widely used as cholesterol-synthesising agents.

True ☐ False ☐

Topic 3 : Microbes in Sewage Treatment

34. Name any two major components in sewage water. ______________

35. Expand STP. ______________

36. Treatment of wastewater is done by ______________ (autotrophic/heterotrophic) microbes which are naturally present in the sewage.

37. Complete the following analogy with respect to sewage treatment.
Primary sewage treatment : Physical : : Secondary sewage treatment : ______________

38. During the primary treatment, floating debris is removed by ________________ and then the grit are removed by ______________.

39. Give the biological term for all the solids that settle down during the primary sewage treatment. ________________

40. The secondary sewage treatment is carried out using the supernatant or effluent of primary sewage treatment.

True ☐ False ☐

41. Arrange the following in the sequence of their requirement during sewage treatment.

I. Large aeration tank II. Anaerobic sludge digester

III. Primary settling tank IV. Secondary settling tank

____

42. Give biological term for the masses of bacteria associated with fungal filaments.

____

43. Flocs are formed in ________________ tanks when effluent is constantly agitated in ________________ environment.

44. What does the picture given below represents? (Primary treatment/Secondary treatment)

____

45. How many of the listed components/organisms are found in secondary sewage treatment?

Primary sludge, Flocs, Activated sludge, Grit

____

46. Expand BOD. ________________

47. During secondary sewage treatment, BOD of effluent reduces due to the consumption of <u>CO_2</u> by the flocs.

(Rewrite the statement by changing the underlined word.)

____

48. BOD refers to the amount of the _______________ consumed if all the organic matters in 1 L water were _______________ by bacteria. The sewage water is treated till the BOD is _____________.

49. The sediment formed by bacterial flocs in secondary settling tank is called

____

50. Digestion of primary sludge by anaerobic bacteria in anaerobic sludge digester form biogas.

True ☐ False ☐

51. The effluent of secondary treatment plant is released into _______________ (aeration tank/natural water body).

52. Name the plans initiated by ministry of environment and forests to save major rivers of India from pollution. _______________

53. The greater BOD value indicates _______________(unpolluted/polluted) water.

Topic 4 : Microbes in Production of Biogas

54. Biogas is a mixture of gases such as methane, _______________ and carbon dioxide.

55. Microbes produce different gases during _______________. The type of gas produced depends upon the microbe and _______________ they utilise. Fermentation of dough produce _____________ gas whereas anaerobic growth of bacteria on cellulosic material produce _______________ gas in large amount.

56. Read the following statements and identify the group of bacteria.

(i) Methane producing bacteria.

(ii) Helps in the breakdown of cellulose in the rumen of cattle.

____

57. Choose the odd one out with respect to the methanogens.

Anaerobic sludge, Hydrothermal vents, Human stomach, Cattle's excreta

____

58. Biogas is commonly called _______________ .

59. The biogas plant consists of a concrete tank in which _______________ is collected.

60. Refer to the diagram of typical biogas plant and identify the parts labelled as *A*, *B* and *C*.

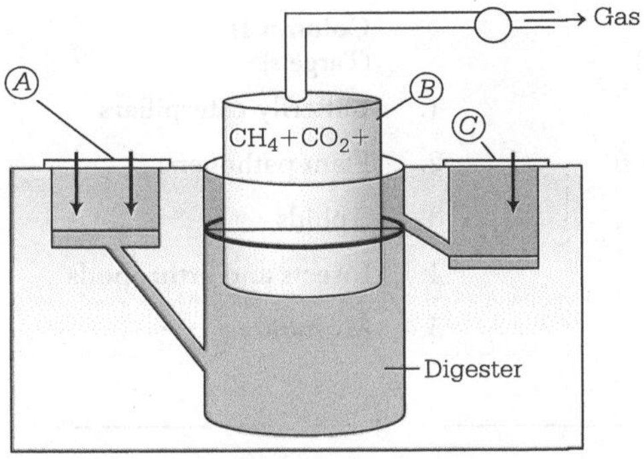

A. ____________________

B. ____________________

C. ____________________

61. The spent slurry removed from biogas plant is used as ____________________ (fertiliser/cattle feed).

62. Arrange the following events in biogas production in the order of their occurrence.

I. Methanogenesis, II. Solubilisation, III. Acidogenesis

____

63. Complete the flowchart depicting the production of biogas in a biogas plant.

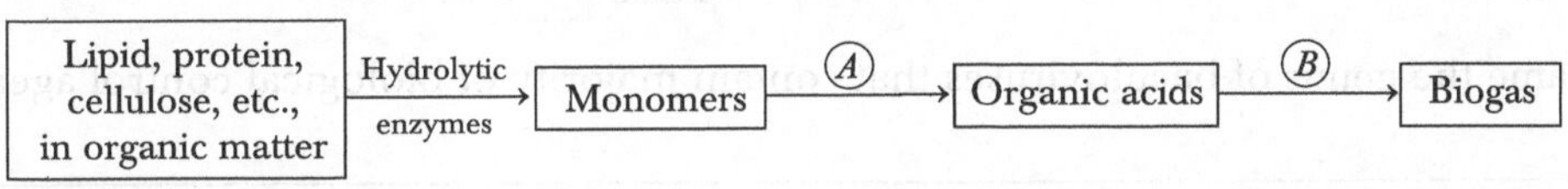

A. ____________________ B. ____________________

64. Expand IARI. ____________________

65. Expand KVIC. ____________________

66. IARI and KVIC contributed in the production of ____________________.

Topic 5 : Microbes as Biocontrol Agents

67. Give the scientific term for the use of biological methods for controlling plant diseases and pests. ____________________

68. Complete the following analogy.
Organic farming : Biocontrol method : : ____________________ : Chemical method

69. Organic farmers believes that eradication of pests affect the survival of beneficial predatory and parasitic insects.

True ☐ False ☐

70. How many of the listed factors are required to be known about pests and predators to develop appropriate means of biocontrol?

Type of pest and predator in field, Life cycles, Feeding habits, Habitat

____

71. Match the following columns.

	Column I (Biocontrol agents)		Column II (Targets)
A.	Ladybird	1.	Butterfly caterpillars
B.	*Bacillus thuringiensis*	2.	Plant pathogens
C.	Dragonflies	3.	Aphids
D.	*Trichoderma*	4.	Insects and arthropods
E.	Baculoviruses	5.	Mosquitoes

A. ________________ B. ________________ C. ________________

D. ________________ E. ________________

72. *Bt* spores are mixed with water and sprayed on to vulnerable plants such as ________________________. When eaten by insect larvae, *Bt* releases ________________________ and the larvae get killed. Plants containing ________________ are insect pests resistant, e.g. *Bt* cotton.

73. *Trichoderma* species are parasitic fungi in the roots of plants.

True ☐ False ☐

74. Name the genus of baculoviruses that contain majority of biological control agents.

____

75. Choose the odd one out with respect to baculoviruses.

Species-specific, Broad spectrum insecticides, Pathogenic organisms, Do not affect non-target insects

____

76. Expand IPM. ____

77. Match the following columns.

	Column I (Biocontrol agents)		Column II (Targets)
A.	*Cactoblastis cactorum*	1.	Citrus rust mite
B.	*Hirsutella thompsonic*	2.	*Leptinotarsa decemlineata*
C.	*Rodiola cardinalis*	3.	Sugarcane leaf hopper
D.	*Cyrtorhinus mundulus*	4.	*Opuntia*
E.	*Beauveria bassiana*	5.	*Icerya purchasi*

A. ________________ B. ________________

C. ________________ D. ________________

E. ________________

Topic 6 : Microbes as Biofertilisers

78. Give the scientific term for the organisms that enrich the nutrient quality of the soil.

____

79. Choose the odd one out with respect to the source of biofertilisers.

Bacteria, Fungi, Penicillin, Cyanobacteria

____

80. Complete the following analogy.

Symbiotic nitrogen-fixing bacteria : *Rhizobium* : : Free-living nitrogen-fixing bacteria :

____

81. *Frankia* is a nitrogen-fixing fungi for non-leguminous plants like *Alder*, barberry, etc. (*Rewrite the statement by changing the underlined word.*)

____

82. Give the biological term for the symbiotic association of fungi with plants.

____

83. Members of genus–*Glomus* do not form mycorrhiza.

True ☐ False ☐

84. In mycorrhizal association, fungal symbiont absorbs ______________ from soil and passes it to associated plant.

85. Do mycorrhizae associated plants show resistance to root borne pathogens and tolerance to salinity and drought?

Yes ☐ No ☐

86. Cyanobacteria are ______________ (autotrophic/heterotrophic) microbes.

87. How many of the listed species are symbiotic nitrogen-fixing cyanobacteria?

Nostoc, Anabaena, Rhodospirillum, Oscillatoria, Clostridium

____

88. Cyanobacteria serve as an important biofertiliser in ______________ (paddy field/wheat plants).

89. Name the cyanobacteria found in sugarcane and maize fields.

____

Answers

1. Bacteria
2. *A*–Rod-shaped bacteria
   B–Spherical-shaped bacteria
3. Obligate
4. *B* is odd one out, as it is a rod-shaped bacterium showing flagella whereas the rest three are viruses, *A* is bacteriophage, *C* is adenovirus and *D* is tobacco mosaic virus.
5. False. The colonies of microbes grown on nutritive media can be seen with the naked eyes.
6. Bacteria, fungi
7. Tobacco mosaic virus
8. Adenovirus
9. Lactic Acid Bacteria
10. Lactic acid, casein
11. Vitamin–B_{12}
12. A-3, B-1, C-4, D-2
13. CO_2 gas, fermentation
14. Toddy
15. False. The particular flavour of Roquefort cheese is imparted by a fungi, *Penicillium roqueforti.*
16. Fermentors
17. Beer and Whisky
18. Ethanol
19. *A*–H_2O, *B*–$C_6H_{12}O_6$ (fructose), *C*–Zymase, *D*– CO_2
20. Wine is odd one out, as it is produced without distillation whereas the other three are produced by the distillation of fermented broth.
21. *A-Hordeum vulgare* (barely)
    B–Molasses of sugarcane
    C–Brandy
22. Antibiotics
23. Penicillin, *Staphylococci*
24. Alexander Fleming, Ernst Chain and Howard Florey.
25. A-4, B-1, C-5, D-3, E-2
26. 5(Tetracycline, kanamycin, chloramphenicol, Rifampicin and Ampicillin). Penicillin and streptomycin are narrow spectrum antibiotics.
27. (i) Streptomycin
    (ii) Azithromycin
    (iii) Erythromycin
    (iv) Rifampin
28. Fungi
29. A-2, B-4, C-1, D-3
30. *A- Candida lipolytica*
    B- Clearing of bottled fruit juices
    C- Streptokinase
    D- Aspergillus niger
    E- Aspergillus niger
31. Cyclosporin-A
32. Yeast, *Monascus purpureus*
33. False. Statins are widely used as blood cholesterol lowering agents as the competitively inhibit cholesterol-synthesising enzymes.
34. Organic waste and microbes.
35. Sewage Treatment Plant
36. Heterotrophic
37. Biological
38. Sequential filtration, sedimentation
39. Primary sludge
40. True
41. III → I → IV → II
42. Flocs
43. Aeration, aerobic
44. Secondary treatment
45. 2 (Flocs and activated sludge). Primary sludge and grit are found in primary sewage treatment.
46. Biochemical Oxygen Demand
47. Organic matter.
48. Oxygen, oxidised, reduced
49. Activated sludge
50. False. Digestion of activated sludge by anaerobic bacteria in anaerobic sludge digester form biogas.
51. Natural water body
52. Ganga Action Plan and Yamuna Action Plan
53. Polluted
54. Hydrogen sulphide
55. Growth and metabolism, organic substrate, CO_2, methane.
56. Methanogens, e.g. *Methanobacterium.*
57. Human stomach is odd one out as methanogens are not found in it while the rest three contain methanogens.
58. Gobar gas
59. Biowaste
60. *A*–Dung+water
    B–Gas holder
    C–Sludge
61. Fertiliser
62. II → III → I
63. *A*–Fermentative microbes
    B–Methanogens
64. Indian Agricultural Research Institute
65. Khadi and Village Industries Commission
66. Biogas
67. Biocontrol
68. Conventional farming
69. True
70. 4 (All of them)
71. A-3, B-1, C-5, D-2, E-4
72. *Brassica* and fruit trees, toxins, *Bt* genes.
73. False. *Trichoderma* species are free-living fungi in the roots of plants.
74. *Nucleopolyhedrovirus*
75. Broad spectrum insecticide is the odd one out as baculoviruses are narrow spectrum insecticides.
76. Integrated Pest Management
77. A-4, B-1, C-5, D-3, E-2
78. Biofertilisers
79. Penicillin is odd one out, as it is an antibiotic while the rest three are the main sources of biofertilisers.
80. *Azotobacter*
81. Actinomycetes
82. Mycorrhiza
83. False. Many members of genus-*Glomus* form mycorrhiza.
84. Phosphorus and sugar
85. Yes
86. Autotrophic
87. 3 (*Nostoc, Anabaena* and *Oscillatoria*) *Rhodospirillum* and *Clostridium* are free-living nitrogen-fixing bacteria.
88. Paddy field
89. *Cylindrospermum licheniformis*

CHAPTER ELEVEN

BIOTECHNOLOGY : PRINCIPLES AND PROCESSES

Topic 1 : Principles of Biotechnology and Restriction Enzymes

1. Controlled multiplication of live organisms for deriving products of human welfare is associated with the field of ________ (biochemistry/biotechnology).

2. How many processes mentioned below are associated with genetic manipulation?

In Vitro Fertilisation (IVF), Gene therapy, Novel gene synthesis, Translocation, Conjugation

   ____

3. Expand EFB . ________________________

4. In genetic engineering process, chemistry of genetic material is altered to induce phenotypic change in host.

   True ☐ False ☐

5. Fill in the blanks.

   (i) ________________ as a part of biotechnology refers to elimination of all living organisms particularly microbial contamination to enable growth of required organism.

   (ii) ________________ engineering involves maintenance of sterile ambience to enable growth of desired cell.

6. Asexual reproductive mechanism provides opportunities for variations and new genetic combinations.

True ☐ False ☐

7. Sexual reproduction preserves the variation which is essential for genetic engineering processes.
(*Rewrite the statement by changing the underlined word.*)

____

8. Traditional hybridisation though convenient is not preferred for creating alterations in organisms selected for breeding due to inclusion and multiplication of undesired genes.

True ☐ False ☐

9. Give scientific term for the following statements.
(i) The technique of altering a host's genome by introducing foreign DNA.

____

(ii) Production of exact copies of a particular gene or DNA sequence.

____

10. A piece of an alien DNA may multiply and be inherited along with the host DNA
(i) When an alien DNA becomes a part of host's genome?
(ii) When all palindromic sites are cleaved by restriction enzymes. (*Choose correct one.*)

____

11. For multiplication in a host cell, an alien piece of DNA should integrate itself with _______________ sequence of the vector.

12. *ori* is the _______________ (stop codon/ initiating sequence for replication) in a host cell.

13. A foreign DNA linked with *ori* can replicate itself in any host cell.

True ☐ False ☐

14. Give a biological term for following statement.
An autonomously replicating, circular, extrachromosomal DNA.

____

15. Plasmid used to construct the first *r*DNA was isolated from _______________ (*Escherichia coli /Salmonella typhi*).

16. _______________ constructed the first artificial recombinant DNA by linking an _______________ gene with _______________.

17. Complete the following analogy given below.
    Restriction enzymes : Molecular scissors :: Plasmid : ________________________

18. DNA can be cut at specific sites in RDT with DNase enzyme.
    (*Rewrite the statement by changing the underlined word.*)

    ____

19. During construction of a recombinant DNA technology, the linking of antibiotic resistance gene with plasmid is done using *Taq* polymerase.

    True ☐ False ☐

20. Plasmids are used as ________________ to deliver an alien piece of DNA into the host organism.

21. The enzyme DNA gyrase, acts on cut DNA molecules and joins their ends.
    (*Rewrite the statement by changing the underlined word.*)

    ____

22. Recombinant DNA when transferred into *Agrobacterium* , a bacterium closely related to *Salmonella*, could replicate using the new hosts's reverse transcriptase enzyme and make multiple copies.

    True ☐ False ☐

23. Father of genetic engineering is ____________________ .

24. To proceed with the process of genetic modification, first the desired DNA is amplified.
    (*Rewrite the statement by changing the underlined word* .)

    ____

25. The most common host used in RDT for maintaining the introduced DNA is ______________ (*E. coli / Salmonella*) .

26. Among the three basic steps involved in genetically modifying an organism, identify the parts labelled as *B* and *C*.

    Identification of DNA with desirable genes
    ↓
    B B. ______________________________
    ↓
    C C. ______________________________

27. Recombinant DNA technology can be accomplished only in the presence of certain key tools. How many of these tools are mentioned in the box given below?

Microscope, Restriction enzymes, Host cells, Polymerase, Vectors, Topoisomerase, Ligases

____

28. Restriction enzymes are lysing enzymes used in RDT.

True ☐ False ☐

29. What is common to scientists mentioned in the box given below?

Nathan, Arber and Smith

____

30. First restriction enzyme with endonuclease activity was _______________.(*Pst* I/ *Hind* II)

31. More than 150 restriction enzymes have been isolated from over 230 bacterial strains. (*Rewrite the statement by changing the underlined word.*)

____

32. Restriction enzymes are a part of defence against bacteriophages in bacteria, which cleave the viral DNA. This system is known as

____

33. The bacterial DNA remains protected from restriction enzymes because they are adenylated.
(*Rewrite the statement by changing the underlined word.*)

____

34. While naming restriction enzymes, the first letter of the name comes from the species of host organism.

True ☐ False ☐

35. Restriction enzymes break _______________ linkages.

36. How many classes of restriction enzymes are known?

____

37. Complete the following analogy given below.
Type I restriction enzyme : *Eco*k I :: Type II restriction enzyme : _______________

38. Pick the odd one out.

Eco RI, *Hind* III, PuC, *Bam*HI, *Sma*I

____

39. Endonucleases are the class of enzymes that break DNA at specific sites from within.

True ☐ False ☐

40. Complete the following analogy given below.
Hind III : *Haemophilus influenzae* :: _______________ : *Bacillus amyloliquefaciens*

41. Enzymes that remove nucleotide one at a time from the ends of DNA molecules are modifying enzymes.

True ☐ False ☐

42. Source of *Eco*RI is _______________ (*E. coli* RI / *E. coli* RY13).

43. Complete the following analogy given below with respect to *EcoRI.*

R : strain :: *co* : ___________________________

44. In *Eco*RI, I stand for the <u>number</u> of enzyme isolated from a particular bacterial strain. (*Rewrite the statement by changing the underlined word.*)

____

45. Some endonucleases are highly specific and operate at specific sites DNA called _____________ .

46. Complete the following analogy given below.

Endonuclease : DNA cuts at palindromic nucleotide sequence : : DNA ligase : _____________________ .

47. Match the following columns.

	Column I (Restriction enzymes)		Column II (Derived microorganisms)
A.	*Hind* III	1.	*Providencia stuartii*
B.	*Pst* I	2.	*Serratia marcescens*
C.	*Pvu* I	3.	*Haemophilus influenzae*
D.	*Sma* I	4.	*Proteus vulgaris*

A. _____________ B. _____________ C. _____________ D. _____________

48. Ligase enzyme mediates the formation of phophodiester bonds between adjacent ______________________ termini, thereby joining the nicks in double-stranded DNA.

49. Restriction sites are palindromic nucleotide sequences.

True ☐ False ☐

50. Each restriction endonuclease functions by
(i) isolating the length of a DNA sequence.
(ii) inspecting the length of DNA sequence.
(*Choose correct one.*)

____

51. Restriction enzymes cut each of the two strands of the double helix at specific points in their ___________________________ backbone.

52. The recognition sequence of restriction enzymes reads the same when read in _______________ direction on either of the strands.

53. Read the sequence of two separate DNA strands
(i) 5′- GTCGAC-3′ (ii) 5′-GAATTC-3′.
What is common to both these sequences?

____

54. In the figure given below

(i) Identify the labels *A-C*.

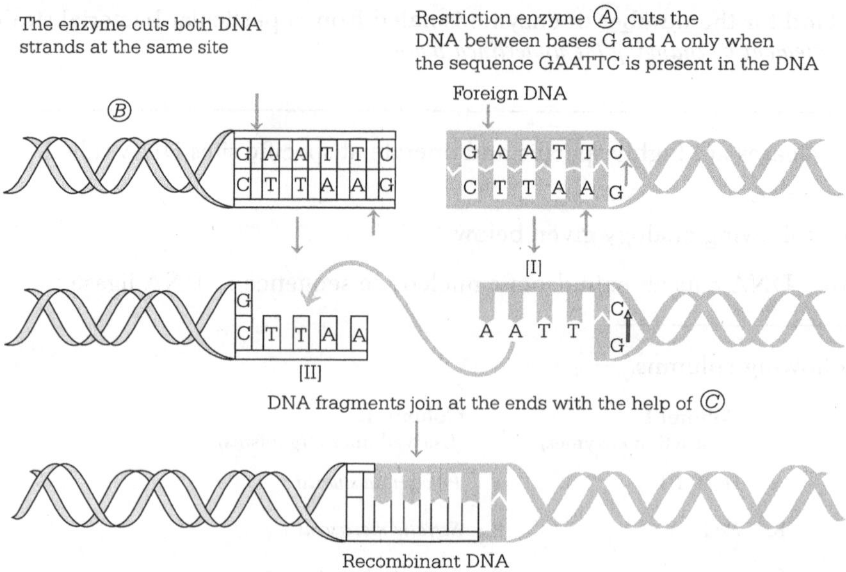

A. ____________

B. ____________

C. ____________

(ii) What kinds of ends are formed at I and II? ____________

55. Identify the parts labelled as *A, B* and *C* in the figure given below.

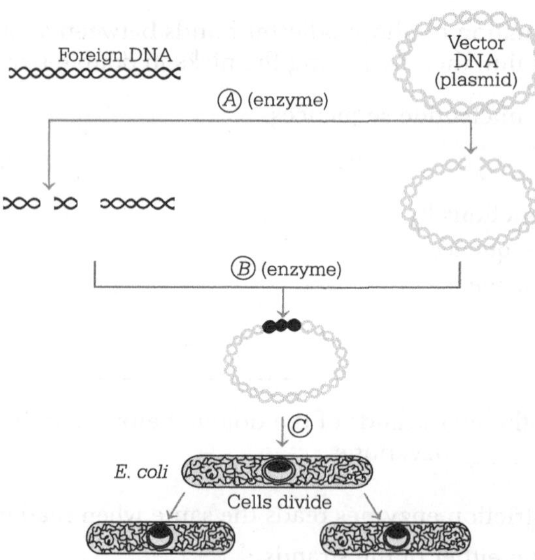

A. ____________

B. ____________

C. ____________

56. Restriction enzymes cut the strand of DNA a little away in the centre of the palindrome sites, between the different bases on the opposite strands.

True ☐ False ☐

57. Fill in the blanks.

(i) *Eco*RI cuts its palindromic sequence which produces overhangs called __________ ends.

(ii) Sticky ends form ____________ bonds with their complementary cut strands.

58. The fragments obtained from vector and host DNA cut by the same restriction enzyme will yield ____________, which facilitate the action of ____________.

59. Why are restriction endonucleases used in genetic engineering processes to

(i) form recombinant molecules DNA?

(ii) generate palindromic sites?
(*Choose correct one*)

____

60. The resultant DNA fragments have the same kind of sticky ends when cut by the ________________________ restriction enzyme.

61. Which enzyme is called as molecular glue?

____

62. Give a scientific term for the following statement.
The process of insertion of *r*DNA in bacterial genome.

____

63. The type of sticky ends of DNA segment produced when restriction enzymes cut in the centre of recognition sequence.

True ☐ False ☐

64. Will a vector cut by restriction enzyme 1 and a source DNA cut by restriction enzyme 2 produce a recombinant molecule?

Yes ☐ No ☐

65. When a restriction enzyme recognises its specific sequence in a given DNA segment?

(i) It binds to it and generate fragments.

(ii) Its first methytate the segment first, then cleave it.
(*Choose correct one.*)

____

66. Give biological term for following statement.

The technique to separate DNA fragments *in vitro* using natural polymer extracted from sea weeds.

____

67. Study the figure given below and answer the following questions.

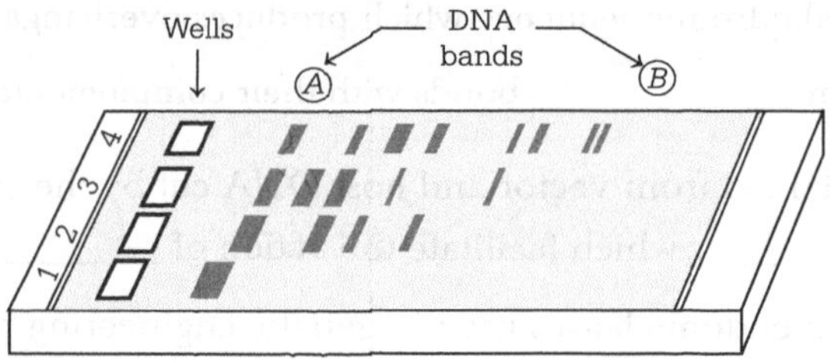

(i) What does the figure respresent? ______________________

(ii) Identify the type of DNA bands present at *A* and *B*.

A.______________ B.______________

68. DNA are positively charged molecules.

True ☐ False ☐

69. Under an electric field, DNA fragments will move towards ______________.

70. The matrix during electrophoretic segregation of DNA fragments is ______________. (agarose/gelatin)

71. The DNA fragments separate based on their size by sieving effect.

True ☐ False ☐

72. The dye used to stain separated fragments of DNA is ______________. (aniline blue/ethidium bromide).

73. Apart from EBTR, exposure to ______________ is required for visualisation of separated DNA fragments appears.

Topic 2 : Cloning Vectors and Competent Host

74. In recombinant DNA technique, the term vector refers to a ______________.

75. How many of the following are cloning vectors used in RDT?

Plasmids, Bacteriophages, *Bt* cotton, Cosmids, YAC, BAC, *Sal* I, Genome, Transposons

____

76. A vector should have a high copy number.

True ☐ False ☐

77. Apart from helping in linking foreign DNA, vectors can also help in ______________ of recombinant cells.

78. Bacteriophages may have 15-100 copies per cell.
(*Rewrite the statement by changing the underlined word.*)

____

79. __________ sequence enables a vector to replicate autonomously and generate multiple copies of itself.

80. An ampicillin resistant gene in a vector is called __________.

81. Identification and elimination of non-transformed cells is a function of *ori*.
(*Rewrite the statement by changing the underlined word.*)

____

82. Normal *E. coli* cells carry resistance against ampicillin and kanamycin.

True ☐ False ☐

83. The vector selected for RDT should have __________ recognition site.

84. The presence of multiple cloning sites generates several DNA fragments which complicate

____

85. The ligation of alien DNA is carried out at restriction site carrying __________ (*ori* /an antibiotic resistance gene).

86. pBR322 and pUC8 are artificial cloning vectors based on bacteriophages.

True ☐ False ☐

87. The recombinant plasmid (pBR322) after losing tetracycline resistance can still be selected from non-recombinants by plating transformants on __________ containing media.

88. In blue-white screening, a fluorescent substrate is used to differentiate recombinant cells.
(*Rewrite the statement by changing the underlined word.*)

____

89. According to blue-white screening method, recombinants undergo inactivation of enzyme __________ (β-galactosidase/ transacetylase).

90. Give a scientific term for the following statement.
The inactivation of a gene by insertion of DNA sequence within it.

____

91. Pick the odd one out.

pUC18, *Bam* HI, YAC, Plasmids

____

92. Plasmids are units of DNA, which can move from one DNA molecule to another.
(*Rewrite the statement by changing the underlined word.*)

____

93. The diagram given below and answer the following questions.

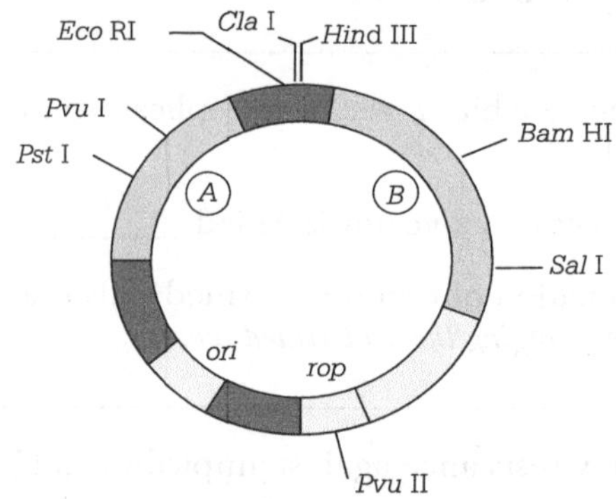

(i) The figure represents a ________________ (cosmid/plasmid) vector.

(ii) Name this vector. ________________

(iii) This vector has two selectable markers *A* and *B*, whose function is *C*. Identify *A*, *B* and *C*.

A.____________ B.____________ C.____________

(iv) To cut at *Bam* HI restriction site, the recognition sequence at 5′-3′ should be read as

____

94. If a DNA fragment is inserted using restriction enzyme *Pst* I or *Puv* I in pBR322, the plasmid will loose ampicillin resistance.

True ☐ False ☐

95. In pBR322, what do B and R represent? ________________ .

96. If the size of gene of interest is more than 10Kb which vector would suit best?

____

97. What is a common feature between M13 and λ-phage?

____

98. A cosmid is constructed using A+B. What are A and B?

____

99. YAC stands for (Expand the term).

____

100. Bacterial artificial chromosome is a vector based on ________________ of *E. coli.*

101. For transforming plant cells________________ (*Agrobacterium tumefaciens*/adenovirus) is used.

102. Plant cells can transform into tumour cells when we insert ________ (B-DNA/ T-DNA) from a known plant pathogen.

103. *Agrobacterium* is a plant pathogen of monocot plants only.

True ☐ False ☐

104. Which part of the *Agrobacterium's* DNA has been modified to be used as a vector for transforming plant cells in genetic engineering?

____

105. ______________ can transform normal cells to cancerous in animals with the help of an enzyme reverse transcriptase.

106. DNA cannot pass through cell membranes because

(i) it is hydrophobic.

(ii) it is hydrophilic.

(*Choose correct one.*)

____

107. Give one word answer for following statement.

The bacterial cell which can uptake a foreign DNA from the surrounding medium.

____

108. Host cell treated with specific concentration of Mg^{2+} becomes competent.

True ☐ False ☐

109. Heat shock (at 42°C) creates ____________________ in the cell membrane of host cells and facilitates uptake of foreign DNA.

110. Pick the odd one out.

Plant tumours, Viruses, Electroporation, Retroviruses

____

111. Complete the following analogy given below.
Animal cells: ____________________ : : Plant cells: Biolistics

112. In biolistics , *r*DNA is directly injected into the nucleus of the host cell.
(*Rewrite the statement by changing the underlined word.*)

____

113. High velocity microparticles delivered into plant cells for transformation are made of ____________________ (gold or tungsten/ lead).

114. For transforming bacterial cells, which method apart from heat shock can be used?

____

115. Disarmed pathogen vectors have been genetically modified to eliminate their ________ .

116. Disarmed pathogen vectors transfer DNA without any risk of infection in host cells.

True ☐ False ☐

Topic 3 : Processes of Recombinant DNA Technology

117. Arrange the following process involved in *r*DNA technology as per their correct sequence

I. Isolation of DNA
II. Ligation of DNA into vector
III. Fragmentation of DNA
IV. Transfer of *r*DNA in host
V. Extraction of desired product
VI. Isolation of desired DNA fragment
VII. Culturing host cells at large scale

____

118. Biochemically the genetic material in most organisms is ________________ (protein/ nucleic acid).

119. Match the following columns.

	Column I (Enzymes)		Column II (Target cells)
A.	Lysozyme	1.	Plant cells
B.	Cellulase	2.	Bacterial cells
C.	Chitinase	3.	Fungal cells

A. __________ B. __________ C. __________

120. When we break open a cell to extract DNA, we also obtain many other constituents. How many of the following can be identified as the same?

RNA, Lipids, Carbon, Proteins, Polysaccharides, Bile, Vitamins

____

121. Complete the following analogy with respect to molecules given.
RNA : ________________ : : Proteins : Protease

122. To remove DNA from the extract, we can use $\underline{CaCl_2}$.
(*Rewrite the statement by changing the underlined word*.)

____

123. Consider the following figure.

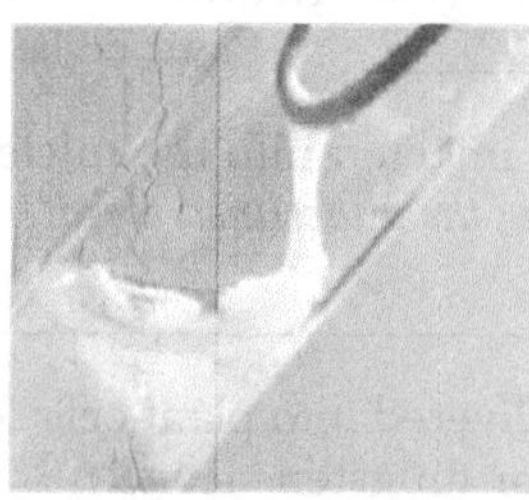

The above figure represents the separation of DNA by method called ____________.

124. Fragmentation of DNA is facilitated by incubation with lyases.

True ☐ False ☐

125. To check the progression of DNA digestion we use <u>SDS-PAGE</u>.
(*Rewrite the statement by changing the underlined word*.)

____

126. Complete the following analogy.
Ligation of vector DNA to host: Ligase: : Amplification of desired DNA: ____________

127. The technique to amplify DNA was developed by Stanley Cohen.

True ☐ False ☐

128. Answer the following questions based on figure given below.
(i) Identify the processes occurring at *A, B* and *C*.

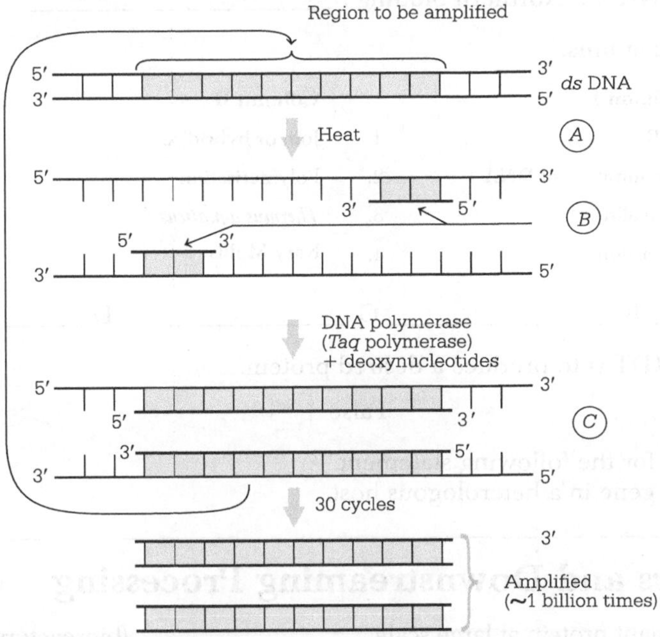

A. ____________

B. ____________

C. ____________

(ii) Complete the following analogy with respect to PCR.
Denaturation : 95°C : : ____________ : 50°C.

129. How many of the following are essential requirements of PCR?

Primers, Ribonuclease, DNA template, Restriction enzymes, Mg^{2+}, *Taq* polymerase, Thermocycler, *d*NTPs

____

130. The *Taq* polymerase is thermostable as it remains active during ____________ (denaturation/extension).

131. *Taq* polymerase is isolated from ____________.

132. Primers are usually ____________ (2-4/10-18) nucleotide long sequences.

133. In PCR, the *Taq* polymerase unwinds the DNA strands.

True ☐ False ☐

134. Using PCR approximately <u>100</u> copies of desired DNA can be obtained.
(*Rewrite the statement by changing the underlined word.*)

____

135. Typically a PCR is run for 60-70 cycles .
(*Rewrite the statement by changing the underlined word.*)

____

136. Molecular diagnosis, DNA fingerprinting, Palaeontology use _______________ to amplify limited amount of DNA.

137. Complete the following analogy given below.
Southern blotting : DNA : : Northern blotting : _______________

138. Match the following columns.

	Column I		Column II
A.	PCR	1.	Join or hybridise
B.	Denaturation of DNA	2.	Polymerisation
C.	Annealing	3.	*Thermus aquaticus*
D.	Extension	4.	Kary Mullis

A. ___________ B. ___________ C. ___________ D. ___________

139. The ultimate aim of RDT is to produce a desired protein.

True ☐ False ☐

140. Give a scientific term for the following statement.
Protein encoded by a gene in a heterologous host.

____

Topic 4 : Bioreactors and Downstreaming Processing

141. To produce a recombinant protein at large scale _______________ (bioreactors / small culture vessels) are used.

142. In _______________ culture system used medium is drained and fresh medium is added.

143. Continuous culture techniques produce less biomass leading to higher yield of desired proteins.

True ☐ False ☐

144. In continuous cultures, the cells are maintained in their _______________ (lag/exponential) phase.

145. In the figure given below which apparatus is shown?

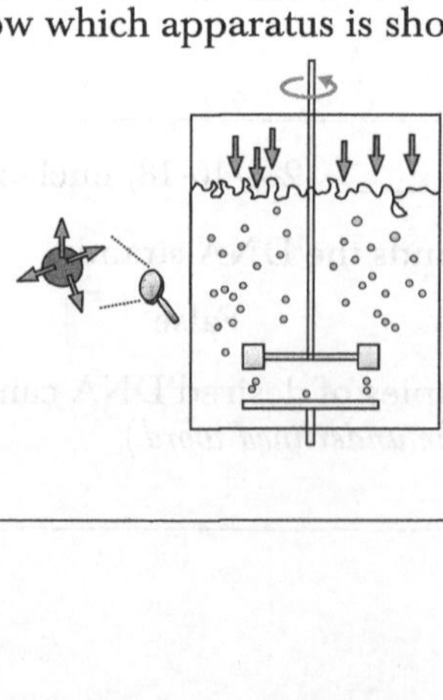

____

146. What is the role of bubbles present the above bioreactor?

____

147. Observe the figure given below and answer in the following questions.
(i) In the figure given below , label all the essential parts.

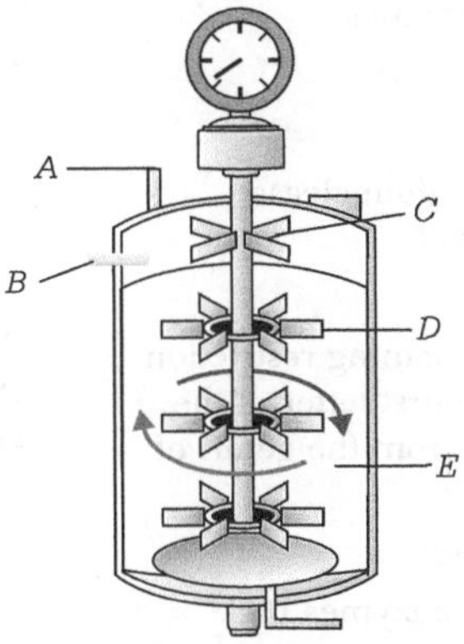

A. ______________
B. ______________
C. ______________
D. ______________
E. ______________

(ii) This type of bioreactor is known as ______________
(iii) In this bioreactor, <u>bubbles</u> facilitate the mixing of contents.
(*Rewrite the statement by changing the underlined word.*)

____

148. Complete the following table given below.

Components of bioreactor	Functions
(A)	Mixes the growth media, breaks any clumps
Oxygen delivery system	(B)
(C)	Prevent contamination in fermentation process
Sampling ports	(D)
(E)	To ensure better dispersal of air
Impellers	(F)

A. ______________ B. ______________
C. ______________ D. ______________
E. ______________ F. ______________

149. Give scientific term.
The process of separation and purification of recombinant proteins.

____

150. To separate the recombinant product from culture broth ______________ (centrifugation /fermentation) is carried out.

151. Before commercially releasing the recombinant product, clinical trials are undertaken to ensure.

____

Answers

1. Biotechnology
2. 3(IVF, gene therapy, novel gene synthesis).
3. European Federation of Biotechnology
4. True
5. (i) Sterilisation
   (ii) Bioprocess
6. False. Sexual reproductive mechanism provides opportunities for variations and new genetic combinations.
7. Permits
8. True
9. (i) Genetic engineering
   (ii) Gene cloning
10. (i) When an alien DNA becomes a part of the host's genome, it multiples and gets inherited along with the host's DNA.
11. *Ori* (origin of replication)
12. Initiating sequence for replication
13. True
14. Plasmid
15. *Salmonella typhi*
16. Stanley Cohen and Herbert Boyer, antibiotic resistance, plasmid
17. Vectors
18. Restriction
19. False. In RDT, the linking of antibiotic resistance gene with plasmid is done using DNA ligase.
20. Vectors
21. DNA ligase
22. False. The bacterium used is *Escherichia coli* which is similar to *Salmonella.*
23. Paul Berg
24. Identified
25. *E. coli*
26. *B* is introduction of desired DNA into host , *C* is transfer of host DNA into its progeny.
27. 5 (restriction enzymes, host cells, polymerase, vectors and ligases).
28. False. Restriction enzymes are cleavage enzymes.
29. Nathan, Arber and Smith together discovered the restriction enzymes.
30. *Hind* II
31. 900
32. Restriction endonuclease enzyme system
33. Methylated
34. False. While naming restriction enzymes, the first letter of the name comes from the genus of host organism.
35. Phosphodiester
36. 3 (restriction enzymes I, II and III)
37. *Eco*RI
38. PuC is odd one out as it is a vector, while others are restriction enzymes.
39. True
40. *Bam* HI
41. False. Enzymes that remove nucleotide one at a time from the ends of DNA molecules are restriction enzymes.
42. *E. coli* RY13
43. Species
44. Order
45. Restriction sites
46. Joins sticky ends
47. A-3 , B-1, C-4, D-2
48. 3′– OH and 5′-phosphate.
49. True
50. (ii) Each restriction endonuclease functions by inspecting the length of a DNA sequence.
51. Sugar-phosphate backbones
52. $5' \rightarrow 3'$
53. Both (i) and (ii) are palindromic sequences, i.e. they read same when read in 5′ - 3′ direction on both strands.
54. (i) *A*–*Eco* RI, *B*–Vector DNA
    C–DNA ligase
    (ii) Sticky ends
55. *A*–Restriction endonuclease
    B–Ligases join foreign DNA to plasmid
    C–Transformation of host cell with *r*DNA
56. False. Restriction enzymes cut the strand of DNA, a little away from the centre of the palindrome sites, but between the same two bases on the opposite strands.
57. (i) Sticky (ii) Hydrogen
58. Sticky ends, DNA ligase.
59. (i) Restriction endonucleases are used in genetic engineering processes to form 'recombinant' molecules of DNA, composed of DNA from different sources.
60. Same
61. DNA ligase
62. Transformation
63. False. Blunt end DNA segments are produced when restriction enzymes cut in the centre of the recognition sequence.
64. No. A recombinant molecule is formed only when source DNA and the vectors are cut by same restriction enzymes.
65. (i) Restriction enzymes after recognising their specific restriction sites bind to it and make a cut generating fragments of varying lengths.
66. Gel electrophoresis
67. (i) Agarose gel electrophoresis for separation of restriction enzyme digested DNA.
    (ii) At *A*- Large size DNA fragments. At *B* - Smallest DNA fragments.
68. False. DNA is a negatively charged molecule.
69. Anode(+ve end)
70. Agarose
71. True
72. Ethidium bromide
73. UV rays
74. Plasmid that transfers DNA into host cell.

75. 6 (plasmids, bacteriophages, cosmids, YAC, BAC and transposons).

76. True. With a high copy number, a foreign DNA can multiply itself equal to the copy number of the vector.

77. Selection

78. Plasmids

79. *Ori*

80. Selectable marker

81. Selectable marker gene

82. False. Normal *E. coli* cells do not carry resistance against ampicillin and kanamycin.

83. Single

84. Gene cloning

85. An antibiotic resistance gene.

86. False. pBR322 and pUC8 are artificial cloning vectors based on plasmids.

87. Ampicillin

88. Chromogenic

89. β-galactosidase

90. Insertional inactivation

91. *Bam* HI is odd one out because it is a restriction enzyme, while others are vectors.

92. Transposons

93. (i) The figure represents a plasmid vector.
(ii) pBR322
(iii) *A* - Ampicillin resistance gene, *B*-Tetracycline resistance gene, *C*-Selection of transformed cells.
(iv) To cut at *Bam* HI, the recognition sequence will be 5′ - GGATCC - 3′.

94. True

95. B and R represent Boliver and Rodriguez, the scientists who developed pBR322.

96. Bacteriophage

97. Both are used as cloning vectors.

98. Cosmids are constructed using (A) cos sites of λ-phage and plasmids (B).

99. Yeast Artificial Chromosome

100. F- factor

101. *Agrobacterium tumefaciens*

102. T-DNA

103. False. *Agrobacterium* is a plant pathogen of dicot plants.

104. Ti plasmid from T-DNA of *Agrobacterium* has been modified to be used as vactor for plants.

105. Retrovirus

106. (ii) DNA is a hydrophilic molecule that is why, it cannot cross cell membranes.

107. Competent cell

108. False. Host cell treated with specific concentration of Ca^{2+} becomes competent.

109. Heat shock creates pore in the cell membrane and facilitates uptake of foreign DNA.

110. Electroporation is the odd one out as it is a direct method of gene transfer while others are indirect method of gene transfer.

111. Microinjection

112. Microinjection

113. Gold or tungsten

114. For transforming bacterial cells, apart from heat shock electroporation can be used.

115. Virulence

116. True

117. The correct sequence is I, III, VI, II, IV, VII and V.

118. Nucleic acid

119. A-2, B-1, C-3

120. 4 (RNA, lipids, proteins and polysaccharides)

121. Ribonuclease

122. Chilled ethanol

123. Spooling

124. False. Fragmentation of DNA is facilitated by incubation with restriction enzymes.

125. Agarose gel electrophoresis

126. PCR

127. False. The technique to amplify DNA was developed by Kary Mullis.

128. (i) *A*–Denaturation, *B*–Annealing of primers, *C*–Extension of primers.
(ii) Annealing

129. 6 (DNA template, primer, thermocycler, Mg^{2+}, *Taq* polymerase and *d*NTPs).

130. Denaturation

131. *Bacillus thermus aquaticus*

132. 10-18

133. False. In PCR, the *Taq* polymerase extends the primers by adding *d*NTPs.

134. 1 billion

135. 30 cycles

136. PCR

137. RNA

138. A-4, B-3, C-1, D-2

139. True

140. Recombinant protein

141. Bioreactors

142. Continuous

143. False. In continuous culturing technique, a larger biomass lead to higher yield of recombinant proteins.

144. Exponential

145. Sparged stirred-tank bioreactor

146. Bubbles increase the oxygen transfer area and facilitate the mixing of culture media.

147. (i) *A*-Acid / Base inlet for pH control, *B*-Steam inlet for sterilisation, *C*-Foam breaker, *D*-Flat bladded impeller, *E*-Culture broth.
(ii) Stirred tank bioreactor
(iii) Impeller

148. *A*- Agitator, *B* - Uninterrupted O_2 supply, *C*- Foam breaker, *D* - To withdrawl of small samples, *E* - Sparger, *F* - Prevents a whirlpool effect that could impede proper mixing.

149. Downstream processing

150. Centrifugation

151. Safety and functionality drugs

CHAPTER TWELVE

BIOTECHNOLOGY AND ITS APPLICATIONS

Topic 1 : Biotechnological Applications in Agriculture

1. How many of the given processes are applications of biotechnology?

Waste treatment, Bioremediation, Genetically modified crops, Conventional hybridisation, Energy generation

____

2. Downstream processing of recombinant products aims to provide the best catalyst, i.e. either an enzyme or a microbe in biotechnological processes.

True ☐ False ☐

3. Read the following statements given below.
   I. Tripling of the food supply.
   II. Increased use of agrochemicals.
   III. Improved crop varieties and yields.

   These statements can be correlated with which major phenomenon in India?

____

4. Due to which factor did the food supply increased during the phase of green revolution?

____

5. Give scientific term for the following statement.

   Plants, bacteria, fungi and animals whose genes have been altered by manipulation.

____

6. An alternate way to maximise the agricultural output without involving the agrochemicals is _______________ .

7. Using conventional breeding methods, we can increase the yield of crops with the existing varieties.

   True ☐ False ☐

8. Most genetically modified plants are produced with the help of <u>*Xanthomonas citri*</u>.
   (*Rewrite the statement by changing the underlined word.*)

   ____

9. As a result of genetically modified crops, chemicals use of an agricultural lands have gone up significantly.

   True ☐ False ☐

10. Genetically modified plants are more tolerant to abiotic stresses than their conventional forms.

    True ☐ False ☐

11. An example of a tailor made crop to supply an enhanced level of vitamin- A is ________________.

12. Golden rice differs from its parental strain by the addition of three ________________ biosynthesis genes.

13. The *Flavr Savr* was made more resistant to ripening and reduce post-harvest loss by adding an antisense gene, which interferes with the production of an enzyme called

    ____

14. *Flavr Savr* is a genetically modified variety of potato.

    True ☐ False ☐

15. Complete the following analogy with respect to GMO crops.

    Golden rice : ________________ :: *Flavr Savr* : Delayed ripening

16. An important objective of biotechnology in agriculture is ________________ (production of insect resistance /to decrease the seed number) in plants.

17. All these plants share a common characteristic very essential to agriculture. Name it.

Bt cotton, *Bt* corn, *Bt* rice, *Bt* tomato and *Bt* potato.

    ____

18. *Bt* toxin is obtained from bacterium called ________________ (*Bacillus thermus aquaticus/Bacillus thuringiensis).*

19. *Bt* toxin has been cloned in a <u>virus</u> and expressed in plants to reduce the addition of agricultural chemicals.
    (*Rewrite the statement by changing the underlined word.*)

    ____

20. An insect belonging to order ________________ (Dermaptera/Lepidoptera) are harmed by the strains *Bacillus thuringiensis.*

21. Genetically modified crops containing *cry* gene were intended to
(i) introduce natural insect resistance in plants.
(ii) increase vitamin content in plants.
(*Choose correct one.*)

____

22. Match the following columns.

	Column I (Orders of class-Insecta)		Column II (Examples of insects)
A.	Dipterans	1.	Armyworm
B.	Coleopterans	2.	Flies
C.	Lepidopterans	3.	Ladybugs

A. ____________ B. ____________ C. ____________

23. Fill in the blanks.
(i) During its a particular growth phase, *B. thuringiensis* forms a special ____________ protein crystal which can kill insects of certain orders, but not their host.
(ii) *Bt* toxin exists as ____________ in host bacteria due to which it remains unaffected.

24. Which factor is responsible for activation of the inactive protoxin as it reaches the target insect's body?

____

25. The *Bt* toxin kills the target insects by <u>inhibiting an essential metabolic pathway</u>. (*Rewrite the statement by changing the underlined words.*)

____

26. *Bt* toxins are not insect group specific.

True ☐ False ☐

27. Complete the following analogy given below.
cry IAc : Cotton bollworms :: ____________ : Corn borer

28. Specific *Bt* toxin genes have been isolated from *Bacillus thuringiensis* and incorporated into how many of the plants mentioned in the box?

Cotton, Chilli, Pea, Corn, Rice, Brinjal, Radish, Potato, Pumpkin and Soyabean

____

29. *Bt* toxin in *B. thuringiensis* is encoded by a ____________ (*cry*/ *CRY*) gene.

30. Identify a potential risk associated with the widespread use of *Bt* technology in agriculture.

____

31. ____________ is a nematode known to infects the roots of tobacco plant and causes great reduction in yield.

32. Which strategy was implemented to prevent parasitic infections in tobacco plants?

___

33. The sequence specific ___ *via* RNAi promises an effective management of agricultural pests.

34. One way, the *ds*RNA can gain entry into eukaryotic cells and cause RNA interference is an infection by ___.

35. RNAi is a method of cellular defence in all prokaryotic organisms.

True ☐ False ☐

36. Silencing with reference to RNAi indicates prevention of <u>transcription of *r*RNA</u> due to its complementary *ds*RNA.
(*Rewrite the statement by changing the underlined words.*)

___

37. Using vectors derived from *E. coli*, nematode specific genes are introduced in the host plant.

True ☐ False ☐

38. RNA interference is a strategy to produce plants with herbicide tolerance.

True ☐ False ☐

39. Complete the following analogy with respect to insect and pest diseases and its host.

Bt cotton : Insect resistance :: ___ : Pest resistance

40. In pest resistance plants, the nematode parasite cannot survive due to

___.

Topic 2 : Biotechnological Applications in Medicine

41. There are atleast ___ different proteins being produced using ___ which serves as therapeutic agents for human beings.

42. How many of the below mentioned substances are recombinant DNA derived therapeutic agents?

Insulin, Growth hormone, α-antitrypsin, Ligase, TPA, Golden rice

___

43. Insulin is used in the treatment of <u>liver damage</u>.

(*Rewrite the statement by changing the underlined words.*)

___

44. Insulin extracted from animal sources other than human body is just as effective in the treatment of concerned disease.

True ☐ False ☐

45. Which organism has been employed for the large scale production of insulin by RDT?

____

46. The trade name of insulin produced using *r*DNA technology is ______________.

47. Insulin derived from animal sources caused ______________ (kidney damage/allergic reactions) in human body.

48. In mammals, insulin is synthesised as a prohormone due to the presence of A-peptide (*Rewrite the statement by changing the underlined word.*)

____

49. Mature insulin consists of four polypeptide chains.

True ☐ False ☐

50. The polypeptide chains of human insulin are linked together by ______________ (covalent bonds/disulphide bridges).

51. Name the biotech company which commercially manufactured the first recombinant human insulin. ______________

52. Arrange the following steps (I–VI) in the order of stages involved in the production of *r*-insulin *via* genetic engineering.

I. Identification of gene required.
II. Joining of chains A and B.
III. Extraction of gene and plasmid.
IV. Gene inserted into plasmid.
V. Transformation in *E. coli*.
VI. Production of *r*- insulin.

____

53. Study the diagram and identify the labels marked *A-D*.

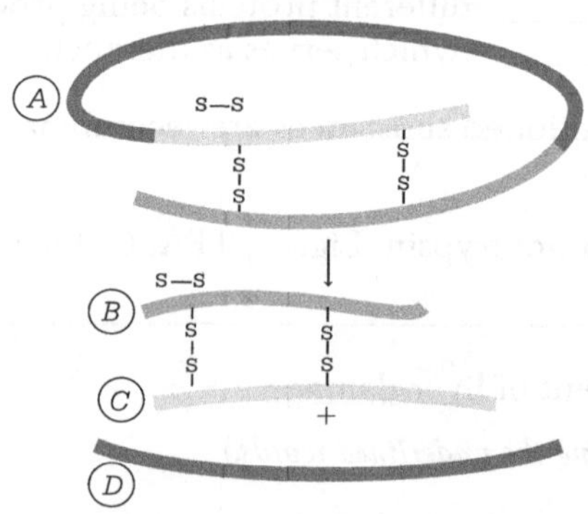

A. ______________

B. ______________

C. ______________

D. ______________

54. In the above diagram, the presence of which 2 chains is essential for the formation of mature insulin? ______________

55. Match the following columns.

Column I (Therapeutics)	Column II (Applications)
A. Growth hormone	1. Pituitary dwarfism
B. α-interferon	2. Haemophilia
C. TPA	3. Cancer
D. Factor VIII	4. Myocardial infarction

A. __________ B. __________ C. __________ D. __________

56. The main challenge faced by companies for the artificial production of mature insulin using genetic engineering techniques was ____________________.

57. If a person is born with a hereditary disorder, he/she is can be cured by ____________________.

58. Gene silencing is the collection of methods to correct gene defects in a child or embryo. (*Rewrite the statement by changing the underlined word.*)

____

59. Correction of genetic disorders involves the repair of non-functional gene.

True ☐ False ☐

60. In somatic cell gene therapy, the expressible gene is introduced into ____________ (bone marrow/sperm or egg cells).

61. When foreign DNA is introduced into sex cells, the process is called as ____________________.

62. In 1990, first clinical gene therapy was provided for ____________________.

63. What is common to all the diseases mentioned in the box given below?

SCID, Cystic fibrosis, Emphysema, Haemophilia-B, Sickle-cell anaemia

____

64. Enzyme ADA is crucial for normal functioning of which system in human body?

____

65. In ____________________ (*ex vivo/ in vivo*) gene therapy, genes from cultured cells are reintroduced into the patient for treatment.

66. In treatment of Adenosine Deaminase (ADA) deficiency disorder, thrombocytes are extracted by doctors for culturing *in vitro.*
(*Rewrite the statement by changing the underlined word.*)

____

67. As an alternate to gene therapy, ADA deficiency can also be curred by transplantation of ____________________ (blood / bone marrow).

68. In some patients, functional ADA is directly injected as a part of somatic gene therapy.

True ☐ False ☐

69. How many of the following can be used as vectors in gene therapy?

Retroviruses, Bone marrow cells, Cosmids

____

70. Why does the ADA deficiency patient require periodic infusion of genetically engineered lymphocytes?

____

71. Gene therapy can be a permanent cure for ADA deficiency disorder if

(i) gene producing ADA is introduced at early embryonic stages.

(ii) bone marrow containing functional lymphocytes is transplanted in adults.

(*Choose correct one.*)

____

72. Urine and serum analysis are ______________ methods of diagnosis.

73. Pick the odd one out.

ELISA, ECG, PCR, RDT

____

74. Give one word for technique of selective amplification of nucleic acids.

____

75. Expand ELISA. ______________________________

76. ______________ is used to detect HIV in patients suspected of AIDS.

77. ELISA technique is based on which principle?

____

78. Give scientific term for the following sentence.

A single strand of DNA/RNA tagged with radioactive molecule used as diagnostic tool.

____

79. Hybridisation of radioactivity tagged DNA/RNA with its *c*DNA can be confirmed using ______________.

Topic 3 : Transgenic Animals

80. Give scientific term for following statement.

Animals whose genome is manipulated to allow expression of a foreign gene.

____

81. 95% of all existing transgenic animals are ______________ (mice/sheep).

82. How many animals have been transgenically created?

Rabbits, Pigs, Sheep, Fish, Crocodile, Frogs

____

83. Novel transgenics are created to study the affect of genes on normal body functions.

True ☐ False ☐

84. Pick the odd one out.

Dolly, *Bt* tobacco, Rosie, Transgenic fish

____

85. Transgenic models exists for how many of the following diseases?

Cancer, Diabetes, Cystic fibrosis, Rheumatoid arthritis, AIDS, Hepatitis-A, Alzheimer's disease

____

86. Match the following columns.

Column I (Transgenic animals)		Column II (Molecular changes introduced)
A. Transgenic cow	1.	Haemoglobin
B. Transgenic goat	2.	Lactoferrin and interferons
C. Transgenic mice	3.	TPA
D. Transgenic pig	4.	Immunoglobulins

A. ________ B. ________ C. ________ D. ________

87. A human protein α-1 antitrypsin is used for treatment of ______________. (emphysema/cystic fibrosis).

88. The first transgenic cow is Dolly. It consists of human alpha-lactalbumin gene.

True ☐ False ☐

89. The milk produced by first transgenic cow was enriched in antithrombin III. (*Rewrite the statement by changing the underlined word.*)

____

90. For what purpose transgenic mice and transgenic monkeys are used during vaccine production?

____

91. Transgenic animals are more sensitive to toxic substances than their non-transgenic forms due to the presence of transgenes.

True ☐ False ☐

92. Give scientific term for following statements.

(i) A vaccine prepared by inserting DNA encoding an antigen. ______________

(ii) Antibodies produced from a single B-lymphocyte clone having monovalent affinity.

____

93. First recombinant vaccine was developed for ________________ .

94. A genetically engineered microbe used in bioremediation of oil spills is ________________ (*Xanthomonas citri* / *Pseudomonas putida*)

Topic 4 : Ethical Issues

95. GEAC stands for ________________

96. <u>CSIR</u> is the organisation responsible for validating the GM research in India. (*Rewrite the statement by changing the underlined word.*)

____

97. Introduction of food/medicines developed through genetic engineering are not much desirable due to

(i) danger of the entry of virus or toxins with transgenic crops.

(ii) low productivity of these products. (*Choose correct one.*)

____

98. Genetically modified organisms can have unpredictable effects on ecosystem.

True ☐ False ☐

99. Give scientific term for following statement.

Intellectual property right granted by government to an individual for novel scientific inventions.

____

100. Patents are granted only for all research works.

True ☐ False ☐

101. India has faced patent related issues for how many of the following products?

Ghee, Turmeric, Rice, Sugar, Neem, Cotton

____

102. In 1997, an American company got patent rights on Basmati rice, which was produced by crossing ________________

103. Approximately ________________ (200,000/27) varieties of rice exist in India alone.

104. Give the scientific term for the following statement.

Illegal use of resources and traditional knowledge without proper authorisation from concerned person.

____

105. ________________ (Indian patents bill/Cartagena protocol) protects the patent rights research and development initiative, etc., of people and protect against biopiracy.

106. The nodal centre for Indian biosafety network is ________________ .

Answers

1. 4 (Waste treatment, bioremediation, genetically modified crops and energy generation).
2. False. Downstream processing aims to purify the organic compound or recombinant proteins obtained by biotechnological processes.
3. Green revolution
4. Use of agrochemicals like insecticides, pesticides, etc.
5. Genetically Modified Organisms (GMOs).
6. Organic farming
7. False. Using conventional breeding methods, we cannot increase the yield of crops with the existing varieties.
8. *Agrobacterium tumefaciens.*
9. False. As a result of genetically modified crops, chemical use of an agricultural lands have gone down significantly in some crops (with little or no change on others).
10. True
11. Golden rice
12. β-carotene
13. Polygalacturonase
14. False. *Flavr Savr* is a genetically modified variety of tomato.
15. Increased vitamin-A content
16. Production of insect resistance.
17. They have *Bt* toxin produced by *cry* gene.
18. *Bacillus thuringiensis*
19. Bacteria
20. Lepidoptera
21. (i) Introduce natural insect resistance in plants.
22. A-2, B-3, C-1
23. (i) Toxic insecticidal protein
    (ii) Protoxin
24. The alkaline pH of the insect's gut is responsible for activation of the inactive protoxin as it reaches the target insect's body.
25. Creating pores in mid gut epithelial cells that cause swelling and lysis of cell.
26. False. *Bt* toxins are insect group specific.
27. *cry* IAb
28. 6 (Cotton, corn, rice, brinjal, potato and soyabean).
29. *cry*
30. The spread of resistance against *Bt* toxin in conventional and modified varieties.
31. *Meloidogyne incognita*
32. RNA interference
33. Gene silencing
34. RNA virus.
35. False. RNAi is a method of cellular defence in all eukaryotic organisms.
36. Translation of *m*RNA.
37. False. *Agrobacterium* vectors are used to introduce nematode specific genes in host plants.
38. False. RNAi is a strategy to produce plants with pest resistance.
39. Tobacco
40. Silencing of specific *m*RNA
41. 30, genetic engineering
42. 4 (Insulin, growth hormone, α-anti-tyrpsin and TPA).
43. Diabetes mellitus
44. False. Insulin extracted from animal source is not as effective as that secreted in human body.
45. *Escherichia coli*
46. Humulin
47. Allergic reactions
48. C-peptide
49. False. Mature insulin consists of two polypeptide chains.
50. Disulphide bridges
51. Eli Lilly
52. The correct sequence is I, IV, III, V, II, VI.
53. *A*-Proinsulin, *B*-A-peptide, *C*-B-peptide, *D*-Free C-peptide
54. Chains-A and B are essential for the formation of mature insulin.
55. A-1, B-3, C-4, D-2
56. Assembly of both newly synthesised polypeptide chains.
57. Gene therapy
58. Gene therapy
59. False. Correction of genetic disorders involves the replacement of non-functional genes.
60. Bone marrow
61. Germ cell gene therapy
62. ADA deficiency
63. The diseases are all due to genetic disorders.
64. Immune system
65. *Ex vivo*
66. Lymphocytes
67. Bone marrow
68. False. In some patients, functional ADA is directly injected as a part of enzyme replacement therapy.
69. 2 (Retroviruses and bone marrow cells).
70. Because genetically modified lymphocytes have limited lifespan.
71. (i) Gene therapy can be a permanent cure for ADA deficiency disorder if gene producing ADA is introduced at early embryonic stages.
72. Conventional
73. ECG is odd one out. Rest all are molecular diagnostic techniques.
74. Polymerase Chain Reaction
75. Enzyme Linked Immunosorbent Assay.
76. PCR test
77. Antigen-antibody interaction
78. Probe
79. Autoradiography

80. Transgenic animals

81. Mice

82. 4 (Rabbits, pigs, sheep, fish).

83. True

84. *Bt* tobacco is odd one out as it is a transgenic plant. Others are transgenic animals.

85. 4 (Cancer, cystic fibrosis, rheumatoid arthritis and Alzheimer's disease).

86. A-2, B-3, C-4, D-1

87. Emphysema

88. False. The first transgenic cow is Rosie.

89. Human α-lactalbumin

90. Transgenic mice and monkeys both are used to check vaccine safety.

91. True

92. (i) Recombinant vaccine
(ii) Monoclonal antibodies

93. Hepatitis -B

94. *Pseudomonas putida*

95. Genetic Engineering Approval Committee

96. GEAC

97. (i) This is because there is danger of entry of virus or toxins with new transgenic crops.

98. True

99. Biopatent

100. False. Patents are granted only for novel inventions.

101. 3(Turmeric, rice and neem).

102. Indian Basmati with semi-dwarf variety.

103. 200,000

104. Biopiracy

105. Indian patents bill

106. Department of Biotechnology

CHAPTER THIRTEEN

ORGANISMS AND POPULATIONS

Topic 1 : Organism and Its Environment

1. ______________ is a subject which studies the interactions among organisms and between the organism and its environment.
2. Write the four levels of biological organisation.

   ____

3. The variations in the intensity and duration of temperature and precipitation account for the formation of biomes.

   True ☐ False ☐

4. Match the following columns.

Column I (Name of biomes)	Column II (Temperture ranges)
A. Tropical forest	1. – 12-2°C
B. Arctic and Alpine tundra	2. 20-25°C
C. Coniferous forest	3. – 2-22°C
D. Temperate forest	4. 0-12°C

A.______________ B.______________ C. ______________ D. ______________

5. Select the major biomes of India from the box given below .

   Coniferous forest, Tropical rainforest, Deciduous forest, Sea coast, Temperate forest, Alpine tundra, Desert

   ____

6. Find odd one out with respect to abiotic factors of ecosystem.

   Temperature, Water, Soil, Wind

   ____

7. Find odd one out with respect to biotic components of ecosystem.

Pathogens, Predators, Parasites, Light

____

8. Over the years, organisms involved adaptations to optimise its survival and reproduction in its habitat through genetic drift.
(*Rewrite the statement by changing the underlined word.*)

____

9. Each organism has an invariably defined range of conditions that it can __________ to diversity in the resources it utilises and a distinct functional role in the __________, all these together comprise its __________.

10. The average temperature __________ (decreases/increases) progressively from the equator towards poles.

11. Average temperature exceeds 50°C in deep sea thermal vents.
True ☐ False ☐

12. Name any two countries where mango trees cannot grow. __________

13. The __________ (temperature/moisture) does not affect the basal metabolism of living organisms.

14. The abiotic factor which determines the species geographical distribution is __________ (wind/temperature).

15. Give the scientific name for the organisms that can tolerate wide range of temperature.

____

16. The organisms that can tolerate very narrow range of temperature are called

_______________.

17. Life on earth originated in water.
True ☐ False ☐

18. The productivity and distribution of plants is dependent on __________

19. Arrange the following in the increasing order of salinity.

Sea water, Hypersaline lagoons, Inland water

____

20. Identify the water bodies which show the given salinity in parts per thousand.
(i) 30-35 __________ (ii) < 5 __________

21. Give biological term for the organisms which tolerate to a wide range of salinities __________ and organisms which are restricted to a narrow range of salinities __________.

22. Herbs and shrubs in forests cannot perform photosynthesis at high light conditions. (*Rewrite the statement by changing the underlined words.*)

____

23. Complete the following sentence by using the words in the box given below.

Diurnal, Nocturnal, Photoperiod, Photosynthesis, Migratory, Respiratory, Foraging, Hunting

The diurnal and seasonal variations in light intensity and ____________ are the area of animals for ________, reproductive and ________ activities.

24. Radiation below the visible light, i.e. infrared is useful to some organisms. (*Rewrite the statement by changing the underlined words.*)

____

25. Match the following columns.

Column I (Categories of organisms)	Column II (Examples)
A. Stenohaline	1. Salmon
B. Eurythermal	2. Shark
C. Euryhaline	3. *Artemisia*
D. Stenothermal	4. Coconut

A.________ B.________ C. ________ D. ________

26. Soil composition, grain size and aggregation determine the ____________ and ____________ of soil.

27. The nature and properties of soil depend on the ____________, ____________ process and how soil development occurred.

28. Find odd one out with respect to vegetation cover of an area.

pH, Topography, Mineral composition, Types of animal

____

29. Sediment characteristics often determine the type of benthic animals in the soil environment. (*Rewrite the statement by changing the underlined word.*)

____

30. Name the process in which the organism should try to maintain constancy of its internal environment.

____

31. 99% of animals are regulators which are also referred as ectotherms.

True ☐ False ☐

32. ________ are organisms which maintain their body temperature despite of change in temperature of their environment.

33. Reptiles and amphibians are example of ____________ (regulators/conformers).

34. ______________________ regulate their temperature up to certain extent.

35. The success of mammals is largely due to their ability to maintain __________ body temperature and thrive in it.

36. Write biological term for the sentence.
Process of maintaining constant osmotic concentration. ______________________

37. In summer, sweating gives a cooling effect to an animal.
True ☐ False ☐

38. Complete the following analogy given below.
Summer : Sweating : : ____________ : Shivering

39. Plants have a mechanism to maintain internal temperature.
True ☐ False ☐

40. Give two examples of regulators. ______________________

41. Write the biological term used for plants and animals which change their internal environment with the ambient environmental conditions. ______________

42. Thermoregulation is <u>thermodynamically</u> expensive for many organisms.
(*Rewrite the statement by changing the underlined word.*)

____

43. Heat gain or heat loss is a function of ______________________

44. Small animals have ____________ (smaller/larger) surface area relative to their volume.

45. Small animals have to expend ____________ (less/more) energy to generate body heat through metabolism.

46. Very small animals are rarely found in polar regions. (*Give reason*)

____

47. Write the two other alternatives that organism adopt for a short duration stressful condition.

____

48. Name any two small animals which cannot afford thermoregulations.

____

49. Identify the parts labelled as *A, B, C* and *D* in this diagrammatic representation of organismic response.

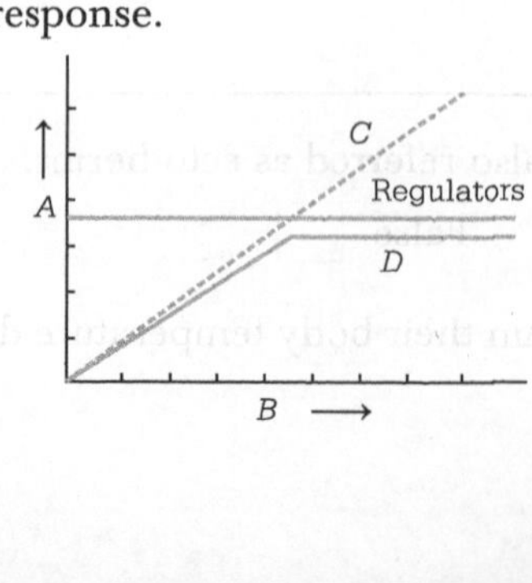

A. ______________

B. ______________

C. ______________

D. ______________

50. Name the very famous National Park in India which hosts the migratory birds from Siberia.

____

51. Thick-walled spore formation during unfavourable condition is seen in bacteria and fungi.

True ☐ False ☐

52. Write a biological term for the sentence.

A state of reduced metabolic activity observed in seeds to tide over the period of stress conditions.

____

53. Write the scientific name of escape in time exhibited by bears during winter.

____

54. ______________ is a escape from summer related heat and desiccation problems.

55. Match the following columns.

Column I (Names of animals)	Column II (Adaptations)
A. Snail	1. Hibernation
B. Zooplankton	2. Aestivation
C. Bear	3. Diapause

A.__________ B.__________ C.__________

56. Name the stage of suspended development in planktons. ______________

57. ______________ is any attribute of an organism which enables the organism to survive and reproduce in its habitat.

58. Write the three major attributes of adaptation found in animals.

____

59. Internal fat oxidation is metabolic reaction in kangaroo rat to meet its fat requirement. (*Rewrite the statement by changing the underlined words.*)

____

60. Write the two beneficiary adaptations of internal fat oxidation.

(i) ______________________________

(ii) ______________________________

61. Thick cuticle on leaf and stomatal arrangement in deep pits is to ______________ (maximise/minimise) water loss through ______________ .

62. Name the plants which carry special photosynthetic pathway in which the stomata remain closed during daytime.

____

63. Complete the following analogy with respect to adaptation in *Opuntia*.

Stem : __________ : : Leaves : __________

64. Photosynthetic function in *Opuntia* is carried over by flattened roots.

True ☐ False ☐

65. Name the rule which assumes mammals from colder climates have shorter ears and limbs to minimise heat loss.

____

66. Seals have a thick layer of fat called ______________ below thin skin and acts as an ______________ to reduce body heat.

67. Write a common term for the following.

Nausea, Fatigue, Heart palpitations

____

68. Biochemical adaptations are seen in organisms in great depths of the ocean to face ______________________ .

69. Desert lizard lacks the physiological condition to deal with high temperature of their habitat, cannot manage to keep their body temperature fairly constant by behavioural means.

True ☐ False ☐

70. Write the two mechanisms by which body compensates for the low oxygen availability.

(i) ______________________________

(ii) ______________________________

71. People living in high altitudes have <u>lower RBCs</u> than the people living in plains. (*Rewrite the statement by changing the underlined words*.)

____

72. Select the odd one out with respect to altitude sickness.

Nausea, Fever, Fatigue, Heart palpitations.

____

73. Write the attribute (behavioural/physiological/morphological) of adaptation for the following .

(i) Lizard barking in seen in cold climates. ______________

(ii) Increased haemoglobin in high altitude people. ______________

(iii) Increased breathing rate due to less oxygen availability. ______________

(iv) *Opuntia's* leaves are reduced to spines. ______________

74. Survival of Archaebacteria in hot deep sea hydrothermal vents is an example of ______________________ (behavioural/ physiological adaptations).

Topic 2 : Population Attributes

75. Name the area of ecology dealing with ecology and population genetics.

____

76. If in a pond, there are 20 lotus plants of last year and through reproduction 8 new plants are added them the birth rate would be ______________.

77. Write the technical name for the ratio of male to female individuals in a population.

____

78. Write the technical term for the following sentence.
Graphical representation of age distribution (per cent individuals of a given age group).

____

79. Write the growth status of the following populations.

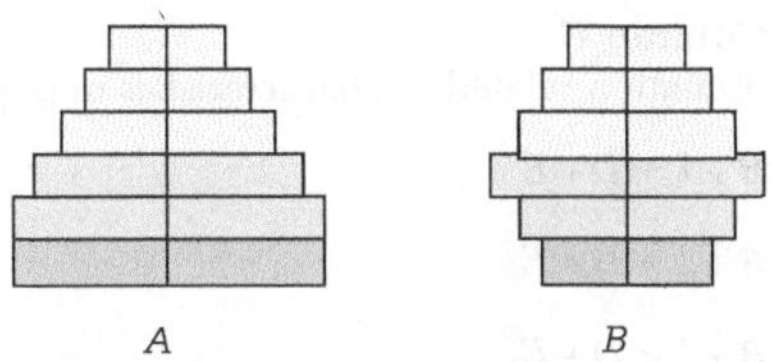

A. ______________

B. ______________

80. Identify the parts labelled as *A, B* and *C* as per their reproductive phase.

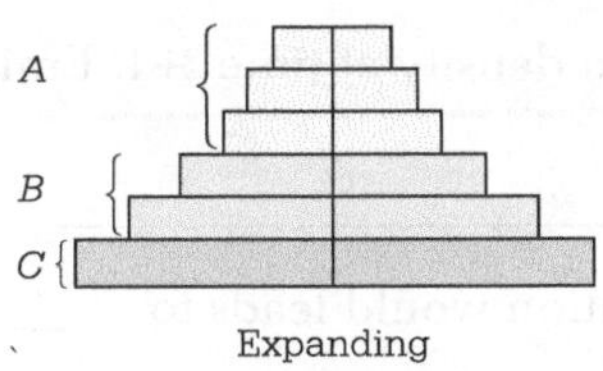

A. ______________

B. ______________

C. ______________

81. Study the following characteristic features of a population.
(i) Proportion of reproductive age group is higher than the pre-reproductive age group.
(ii) Number of post-reproductive individuals are moderate.
(iii) Declining or diminishing population.
Above characters shown indicate which type of age pyramid?

____

82. Name the process by which tiger census done in our national parks and tiger reserves.

____

83. In population ecology, '*N*' refers to ______________

84. Write the technical name for population size.

____

85. The size of a population for any species is a static parameter.

True ☐ False ☐

86. Write four basic processes that fluctuate the population density.

____

87. The number of deaths in a population during a given period is called ________.

88. Complete the flowchart.

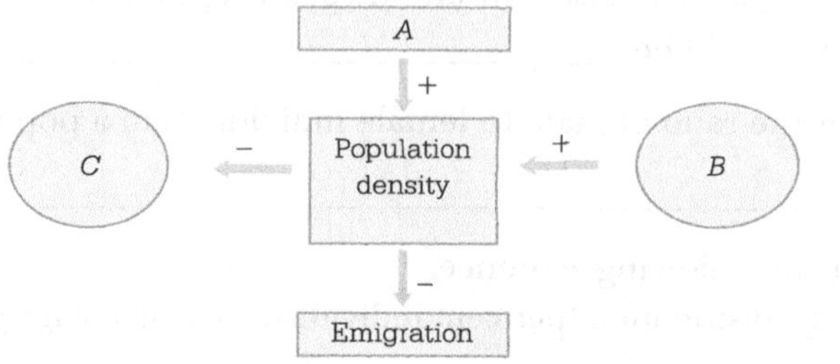

A. ________
B. ________
C. ________

89. Match the following columns if B denotes Natality, I is Immigration, D is Mortality and E is Emigration.

Column I (Characteristics of population)	Column II (Equation related to characteristics of population)
A. Stable population	1. $B+I>D+E$
B. Population is increasing	2. $B+I=D+E$
C. Decreasing population	3. $B+I<D+E$

A.________ B.________ C. ________

90. Write an equation which correlates population density at time $t+1$, Emigration, Immigration, Natality and Mortality.

____

91. Increase in the number of births and immigration would leads to ________ (increase/decrease) of population density.

92. For a new colony, birth rate would contribute more to population growth than the immigration.
(*Rewrite the statement by changing the underlined words.*)

____

93. Write the two models of population growth.

____

94. Match the following columns.

Column I (Population attributes)	Column II (Characteristic features)
A. Emigration	1. Number of births during a given period
B. Mortality + Emigration	2. Number of individuals leaving habitat
C. Natality	3. Increase in population
D. Natality + Immigration	4. Decrease in population

A.________ B.________ C. ________ D. ________

95. Unlimited resources availability leads to impeded growth of a population.
(*Rewrite the statement by changing the underlined word.*)

____

96. Write the mathematical equation for the expression of change in the population size of a population in a unit time period. ____________________

97. Name the parameters used for assessing the impact of biotic or abiotic factor on population growth.

____

98. In 1981, the 'r' value for human population in India was 0.0205.

True ☐ False ☐

99. Write the shape of the curve obtained in a plot between population density and time for exponential growth.

____

100. Write the integral form of exponential growth equation.

____

101. Any species growing ________________ (exponentially/logistically) under unlimited condition can reach enormous population density.

102. A limited resource would lead to ________________ in many individuals.

103. Write biological term for the following sentence.
The limit of a possible number of a population in a given habitat beyond which no further growth possible. ____________________

104. 'K' stands for K-selected species in the Verhulst-Pearl logistic growth equation.

True ☐ False ☐

105. Complete the following paragraph by using the words given below in the box.

Deceleration, Growth, Log, Carrying capacity, Lag, Population density

A population in a habitat with limited resources show initially a ____________ phase, followed by phases of acceleration and ________________ and finally an asymptote, when the population density reaches the________.

106. Carrying capacity of a population is determined by limiting resources.

True ☐ False ☐

107. A plot of N in relation to time 't' when resources are limited is ________________ curve.

108. Write the Verhulst-Pearl logistic growth equation.

____

109. Name the growth model which is considered as a realistic one.

____

110. Identify the growth models given below.

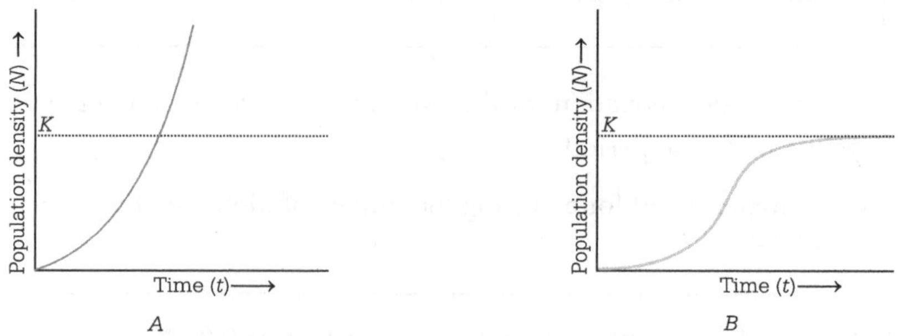

A. ______________________ B. ______________________

111. Darwin fitness has __________ (high/low) *r* value in the habitat in which they live.

112. Write any two organisms which breed only once in their lifetime.

____

113. Select the incorrect pair.

(i) Small number but large-sized offspring–Birds/mammals

(ii) Breeds many times – Mammals

(iii) Small-sized but many offspring – Prey birds/Raptors

____

114. Ecologists suggest that life history traits of organisms have evolved in relation to the __________ imposed by the __________ and __________ components of the habitat.

Topic 3 : Population Interactions

115. Plant species need soil microbes to breakdown __________ matter in soil and to return the __________ nutrients for absorption.

116. Organisms can live in isolation in nature.

True ☐ False ☐

117. Interspecific interactions occur from separation of population of two same species. (*Rewrite the statement by changing the underlined words.*)

____

118. Write the type of interaction assigned by the following symbols.

(i) + ______________________

(ii) 0 ______________________

119. __________ is the type of population interaction in which both the species are harmed.

120. In amensalism, one species is harmed and the other is unaffected.

True ☐ False ☐

121. The interaction where one species is benefitted and the other is neither benefitted nor harmed is called <u>mutualism</u>.
(*Rewrite the statement by changing the underlined word.*)

____

122. Table given below shows the positive interaction (+) and negative interaction – between the species *A* and species *B*. Study the table and complete the missing words *A* to *D*.

Species A	Species B	Interaction
+	+	Ⓐ
–	Ⓑ	Amensalism
Ⓒ	0	Commensalism
+	–	Ⓓ

A.__________ B.__________ C.__________ D.__________

123. Find odd one out with respect to the population interaction.

Predation, Parasitism, Commensalism, Competition.

____

124. Sparrow eating any seed is an example of predation.

True ☐ False ☐

125. Predators act as conduits for __________ transfer across the trophic levels.

126. A new species becomes invasive due to the inavailability of its natural predators.

True ☐ False ☐

127. Name the cactus that became invasive in Australia.

____

128. __________ control methods are based on the ability of predator to control its prey population.

129. Extinction of prey population ultimately leads to the extinction of __________.

130. Proto-cooperation is a defence mechanism to loosen the predation impact.

True ☐ False ☐

131. Name the butterfly which becomes distasteful to its predator bird.

____

132. Write the source of the chemical which protects the monarch butterfly from its predator species.

____

133. Write the scientific name for the insects feeding on plant sap.

____.

134. Write two plants that have thorns as a morphological means of defence.

____

135. Name the scientific name of the weed that produces highly poisonous cardiac glycosides.

____

136. Find odd one out with respect to chemical defence substances found in plants against grazers.

Nicotine, Caffeine, Opium, Chloroquine

____

137. Write any two chemical substances produced by the plants for defence.

____

138. Competition occurs among closely related species competing for limited resources.

True ☐ False ☐

139. In competition, intrinsic rate of increase of one species is lowered due to the presence of another species.

True ☐ False ☐

140. Abingdon tortoise in Galapagos island became extinct due to parasitism.

True ☐ False ☐

141. Match the following columns.

Column I (Interacting organisms)	**Column II** (Interactions)
A. Flamingo and fishes	1. Commensalism
B. Egret and cattle	2. Competition
C. *Nepenthes*	3. Predator

A.______________ B.______________ C. ______________

142. Expansion of distributional range after removal of competing species is called as

____

143. Write the scientific name of the following.

(i) Superior barnacle ________________

(ii) Smaller barnacle ________________

144. Carnivores are more affected than herbivores by competition.

True ☐ False ☐

145. Name the principle which states that two closely related species competing for same resources cannot co-exists.

____

146. Name the mechanism of adapting different times for foraging among two species competing for same resource.

____

147. Match the following columns.

	Column I (Types of principle)		Column II (Scientist's names)
A.	Struggle for existence	1.	MacArthur
B.	Competitive exclusion principle	2.	Darwin
C.	Resource partitioning	3.	Gause

A.__________ B.__________ C. __________

148. Niche overlap indicates two different parasites on the same host.

True ☐ False ☐

149. Parasites evolved special adaptations such as loss of __________, high reproductive capacity and presence of __________ to cling to host.

150. Human liver fluke has __________ (one/two/three) intermediate hosts.

151. Human liver fluke is an annelid.

True ☐ False ☐

152. __________ are the parasites that feed on external surface of the host organism.

153. Find odd one out from the following species.

Lice, Ticks, Copepods, Fluke

____

154. Write the scientific name of the parasitic plant that had lost its chlorophyll and leaves.

____

155. Write the common name for the parasites that live inside the host body.

____

156. Female mosquito is not considered as a parasite. (*Give reason*)

____

157. Endoparasites emphasize more on their reproductive potential.

True ☐ False ☐

158. Select ectoparasites among the following.

Lice, Copepods, *Cuscuta, Taenia, Ascaris,* Liver fluke, Ticks.

____

159. The parasitism in which cuckoo bird lays, its eggs on the crow nest during breeding season is called ______________________ .

160. Identify the interaction between the following species.

(i) Orchid growing on a mango tree. ______________________

(ii) Barnacles growing on the back of a whale. ______________________

161. Match the following columns.

Column I (Parasitic organisms)	Column II (Hosts)
A. Cattle egret	1. Clown fish
B. Orchid	2. Pseudoscorpions
C. Sea anemone	3. Livestock
D. Beetles	4. Mango tree

A.__________ B.__________ C. __________ D. __________

162. The mutualistic relationship between a fungus and photosynthesising algae or cyanobacteria is called ______________ .

163. Name the association between fungi and roots of higher plants.

____

164. Write any two rewards plant offers for pollinators.

____

165. Co-evolution of the mutualists can be found between fig species and wasp species.

True ☐ False ☐

166. Female wasp uses the fig inflorescence as an ______________ site.

167. Give the scientific name for the orchid which employs sexual deceit to get pollination done.

____

168. In Mediterranean orchid, the flower petal resembles the female bee.

True ☐ False ☐

169. The male bee is attracted to the orchid petal assuming it as the female bee due to its resemblance and <u>copulates</u> with the flower.
(*Rewrite the statement by changing the underlined word.*)

____

Answers

1. Ecology
2. Organisms → Populations → Communities → Biomes.
3. True
4. A-2, B-1, C-4, D-3
5. Tropical rainforest, Deciduous forest, Desert, Sea coast.
6. Wind is odd one out. Rest of the options are major abiotic factors.
7. Light is odd one out. Rest of the options are biotic components of a habitat.
8. Natural selection
9. Tolerate, ecological system, niche
10. Decreases
11. False. Average temperature exceeds 100°C in deep sea thermal vents.
12. Canada and Germany
13. Moisture
14. Temperature
15. Eurythermal
16. Stenothermal
17. True
18. Water
19. Inland water < Sea water < Hypersaline lagoons.
20. (i) Sea water, (ii) Inland water
21. Euryhaline, stenohaline
22. Low light
23. Photoperiod, foraging, migratory.
24. UV is harmful
25. A-2, B-3, C-1, D-4
26. Percolation, water holding capacity
27. Climate, weathering.
28. Types of animal is odd one out. Rest of the options are the factors that determine the vegetation in that area.
29. Aquatic
30. Homeostasis
31. False. 99% of animals are conformers which are also referred as ectotherms.
32. Regulators
33. Conformers
34. Partial regulators
35. Constant
36. Osmoregulation
37. True
38. Winter
39. False. Plants do not have a mechanism to maintain internal temperature.
40. All birds and mammals
41. Conformers
42. Energetically
43. Surface area
44. Larger
45. More
46. The small animals have a larger surface area relative to their volume they lose body heat very fast when it is cold outside, then they have to expend much energy to generate body heat through metabolism.
47. Migration and suspension
48. Shrews and humming bird.
49. *A* - Internal level
    B - External level
    C - Conformers
    D - Partial regulators
50. Keolado National Park (Rajasthan)
51. True
52. Dormancy
53. Hibernation
54. Aestivation
55. A-2, B-3, C-1
56. Diapause
57. Adaptation
58. Morphological, physiological and behavioural
59. To meet its water requirement.
60. (i) Water is obtained as a byproduct.
    (ii) Concentrating urine to reduce water amount in excretion.
61. Minimise, transpiration
62. CAM pathway
63. Phylloclade, spines.
64. False. Photosynthetic function in *Opuntia* is carried over by flattened stems.
65. Allen's rule
66. Blubber, insulator
67. Altitude sickness
68. Crushing pressure
69. False. Desert lizard lacks the physiological condition to deal with high temperature of their habitat, but manage to keep their body temperature fairly constant by behavioural means.
70. (i) Increasing RBCs production
    (ii) Increasing breathing rate.
71. Higher RBCs
72. Fever is odd one out, as it is not a symptom of altitude sickness.
73. (i) Behavioural
    (ii) Physiological
    (iii) Physiological
    (iv) Morphological
74. Physiological adaptation
75. Population ecology
76. 8/20 = 0.4 offspring per lotus per year.
77. Sex ratio
78. Age pyramid
79. *A*-Stable, *B*-Declining
80. *A* - Post-reproductive
    B - Reproductive
    C - Pre-reproductive
81. Urn-shaped age pyramid
82. Pug marks and fecal pellets
83. Population density
84. Population density
85. False. The size of a population for any species is not a static parameter.
86. Natality, Mortality, Emigration, Immigration.
87. Mortality
88. *A*-Immigration, *B*-Natality *C*-Mortality
89. A-2, B-1, C-3.
90. $N_{t+1} = N_t + [(B + I) - (D + E)]$
91. Increase
92. Immigration, birth rate.

93. Exponential model and logistic model
94. A-2, B-4, C-1, D-3
95. Unimpeded
96. $\frac{dN}{dt} = (b - d) \times N$
97. 'r' or intrinsic rate of natural increase.
98. True
99. J-shaped curve
100. $N_t = N_0 e^{rt}$
101. Exponentially
102. Competition
103. Carrying capacity
104. False. 'K' stands for carrying capacity in the Verhulst-Pearl logistic growth equation.
105. Growth, deceleration, carrying capacity.
106. True
107. Sigmoid
108. $\frac{dN}{dt} = rN\left(\frac{K - N}{K}\right)$
109. Logistic growth
110. *A*-Exponential growth
B-Logistic growth
111. High
112. Pacific salmon fish, bamboo
113. (iii) is incorrect. It can be corrected as
Small-sized but many offspring oysters and pelagic fishes, etc., produce large number of small-sized offspring.
114. Constraints, biotic, abiotic
115. Organic, inorganic
116. False. Organisms cannot live in isolation in nature.
117. Interaction of population of two different.
118. (i) Beneficial interaction
(ii) Neutral interaction
119. Competition
120. True
121. Commensalism
122. *A*–Mutualism, *B*–0
C – +, *D*–Parasitism
123. Competition is odd one out. Rest of options share a common character that the interacting species live closely together.
124. False. Sparrow eating any type of seed is not an example of predation.
125. Energy
126. True
127. Prickly pear cactus
128. Biological
129. Predators
130. False. Camouflaging is a defence mechanism to loosen the predation impact.
131. Monarch butterfly
132. The butterfly acquired the chemical during its caterpillar stage by feeding on a poisonous weed.
133. Phytophagous
134. *Acacia* and cactus
135. *Calotropis*
136. Chloroquine is odd one out. Rest of the options are chemical substances produced by plant as defence against grazers.
137. Quinine and strychnine
138. True
139. True
140. False. Abingdon tortoise in Galapagos island became extinct due to competition with goats.
141. A-2, B-1, C-3
142. Competitive release
143. (i) *Balanus*, (ii) *Chathamalus*
144. False. Herbivores are more affected than carnivores by competition.
145. Competitive exclusion principle
146. Resource partitioning
147. A-2, B-3, C-1
148. True
149. Digestive system, hooks and sucker.
150. Two
151. False. Human liver fluke is a trematode.
152. Ectoparasites.
153. Fluke is odd one out. Rest of the species are ectoparasites.
154. *Cuscuta*
155. Endoparasites
156. The female mosquito needs blood for reproduction and not for nutrition. It does not bite human for its meal.
157. True
158. Lice, copepods, ticks, *Cuscuta*
159. Brood parasitism
160. (i) Commensalism
(ii) Commensalism
161. A-3, B-4, C-1, D-2
162. Lichen
163. Mycorrhizae
164. Pollen and nectar
165. True
166. Oviposition
167. *Ophrys*
168. True
169. Pseudocopulates

CHAPTER FOURTEEN

ECOSYSTEM

Topic 1 : Ecosystem : Structure and Function, Productivity and Decomposition

1. An ecosystem is a ____________ unit of nature, where living organisms interact among themselves and also with the surrounding ____________. The term ecosystem was coined by ____________ in 1953.

2. Name the two basic categories of ecosystem. ____________

3. Choose the odd one out with respect to the type of ecosystem.

Forest, Estuary, Grassland, Desert

____

4. Give two examples of man-made ecosystem. ____________

5. The physical structure of an ecosystem is formed by the interaction of its plants and animals.
(Rewrite the statement by changing the underlined words.)

____

6. Give the biological term for the vertical distribution of different species occupying different levels in biotic community. ____________

7. In a forest ecosystem, trees occupy ____________ vertical strata, shrubs the second and hurbs and ____________ occupy the bottom layers.

____

8. How many of the listed aspects functions as a unit in an ecosystem?

Productivity, Decomposition, Energy flow, Stratification, Biome, Nutrient cycling

____

9. All the four basic functional components of an ecosystem are found in a pond water having shallow water body.

True ☐ False ☐

10. Complete the flowchart given below.

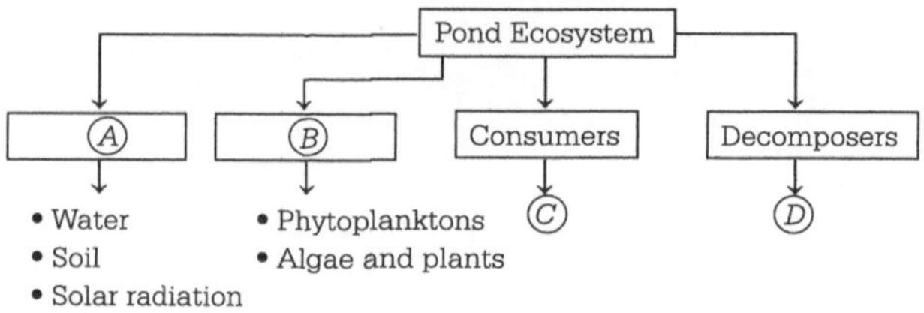

A. ______________ B. ______________ C. ______________ D. ______________

11. Identify the components of pond ecosystem based on the given characteristics/functions.
(i) Name the component which contains all dissolved inorganic and organic substances.

____

(ii) Name the component which converts inorganic into organic material.

____

(iii) Name the component which is especially abundant in the bottom of the pond.

____

12. Name the two processes by which inorganic matter is released from dead remains and made available to autotrophs in a pond ecosystem.

____

13. In a pond ecosystem, there is ______________ movement of energy towards the ______________ trophic levels. It is accompanied by energy dissipation and loss as ______________ to the environment.

14. Match the following columns.

	Column I		Column II
A.	Productivity	1.	Amount of biomass produced per unit area over a time period by plants during photosynthesis.
B.	Primary productivity	2.	Gross primary productivity – Respiration losses.
C.	Secondary productivity	3.	Rate of biomass production.
D.	Net Primary Productivity (NPP)	4.	Rate of new organic matter formation by consumers.

A. ______________ B. ______________ C. ______________ D. ______________

15. The unit of productivity is ______________ $(gm^{-2}\ yr^{-1}/gm^{2}/yr^{-1})$.

16. The available biomass for the consumption of heterotrophs is <u>Gross Primary Productivity (GPP)</u>.
(*Rewrite the statement by changing the underlined words.*)

____

17. How many listed factors can affect the primary productivity of an ecosystem?

Plant species habitat, Environmental factors, Presences of decomposers, Nutrient availability, Photosynthetic capacity of plants

____

18. Complete the following analogy given below with respect to productivity.
Annual NPP of biosphere : 170 billion tons : : Annual NPP of oceans : ________________

19. Primary productivity of different ecosystem varies and oceans are highly productive as they occupy 70% of the surface.
True ☐ False ☐

20. Name the biological process in which an organism breakdown complex organic matter into inorganic substances. ________________

21. The raw material for decomposition is called <u>detritivore</u>.
(*Rewrite the statement by changing the underlined word.*)
____

22. Choose the odd one out with respect to the composition of detritus.

Dead leaves and flowers, Faecal matter, Inorganic substances, Dead remains of animals

____

23. The organisms which carry out the process of decomposition are known as ________ .

24. Match the following columns.

Column I (Steps of decomposition)	Column II (Characteristics)
A. Fragmentation	1. Inorganic nutrients seep into soil.
B. Humification	2. Humus release inorganic nutrients due to the action of microbes.
C. Leaching	3. Breakdown of detritus into smaller particles.
D. Mineralisation	4. Nutrient rich dark-coloured amorphous substance is formed.
E. Catabolism	5. Detritus is broken down by bacterial and fungal enzymes.

A. ________ B. ________ C. ________ D. ________
E. ________

25. Name the detritivore which is referred to as farmer's friend.
____

26. Refer to the given list and answer the questions that follows.

I. Leaching, II. Catabolism, III. Humification, IV. Mineralisation, V. Fragmentation

(i) Arrange the following steps in the order of their occurrence during the process of decomposition. ________________
(ii) How many of these steps operate simultaneously on the detritus? ________________
(iii) How many of these steps occur in the soil? ________________

27. Humus is highly susceptible to microbial action and undergoes decomposition at faster rate.
True ☐ False ☐

28. Being colloidal in nature, ________ (humus/detritus) serves as a reservoir of nutrients.

29. What does the diagram given below represent?

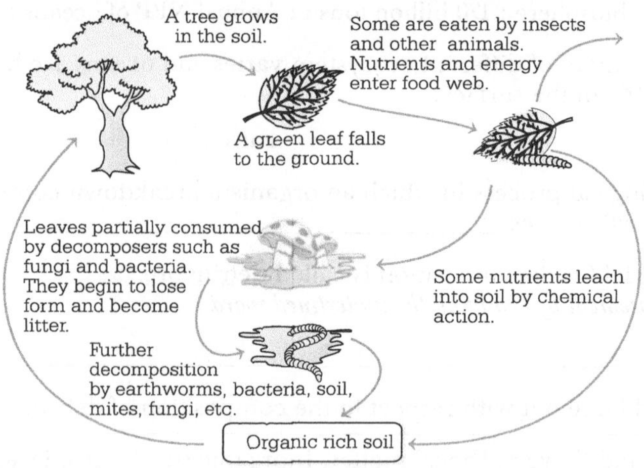

____

30. Decomposition is largely an ____________ requiring process. The rate of decomposition is controlled by ______________________ of detritus and ________________ factors.

31. Complete the following analogy given below with respect to process of decomposition.
Slower decomposition rate : Lignin and chitin in detritus : : Faster decomposition rate:

____

32. Name any two climatic factors that regulate decomposition through their effects on the activities of soil microbes.

____

33. Choose the odd one out with respect to their effects on the rate of decomposition.

Warm environment, Anaerobiosis, Moist environment

____

34. Low temperature inhibits the process of decomposition, resulting in build up of organic material.

True ☐ False ☐

35. The soil with acidic pH and reduced aeration ____________ the process of decomposition.

Topic 2 : Energy Flow and Food Chain

36. Sun is the only source of energy for all ecosystems on earth, except

____

37. Expand PAR. ______________________________

38. What percentage of PAR is captured by plants to sustain the entire living world?

____

39. Energy flow in ecosystem occurs in accordance with first law of thermodynamics because flow of energy is <u>multidirectional</u>.
(*Rewrite the statement by changing the underlined word.*)

____

40. Ecosystems are exempt from the second law of thermodynamics, they do not need a constant supply of energy to synthesise the molecules.

True ☐ False ☐

41. Give the common term for the following words.

Herbaceous and woody plants, Phytoplanktons, Algae

____

42. Give the biological term for the sequence of species through which the food energy pass in a community. ____________________

43. Food chains always start with a ____________ (producers/consumers), an organism that makes food.

44. The energy trapped by plants remains in them forever.

True ☐ False ☐

45. Match the following columns.

Column I	Column II
A. Primary consumers	1. Plant-eating animals
B. Secondary consumers	2. Flesh-eating animals
C. Tertiary consumers	3. Herbivores

A. __________ B. __________ C. __________

46. How many of the listed herbivores are found in terrestrial ecosystem?

Insects, Birds, Mammals, Molluscs

____

47. Complete the following analogy given below with respect to consumers.

Primary carnivores : Secondary consumers : : Secondary carnivores : __________ .

48. Expand GFC. ____________________

49. Identify the correct grazing food chain among (i) and (ii) given below.

(i) Grass → Goat → Man

(ii) Cow → Grass → Goat

____

50. A grazing food chain begins with <u>herbivores</u>.
(*Rewrite the statement by changing the underlined word.*)

____

51. Expand DFC. ____________________

52. Detritus food chain derives energy from ______________ . (sun/organic matter)

53. Name two organisms which to start detritus food chain.

____

54. Read the following statements and identify the type of organisms.

I. Heterotrophic organisms which degrade dead organic matter.

II. Secrete digestive enzymes to breakdown detritus into inorganic materials so as to absorb it.

____

55. The major conduct of energy in aquatic ecosystem is ____________ food chain whereas in a terrestrial ecosystem much larger fraction of energy flow through ____________ food chain.

56. Give the biological term for the natural interconnections of food chains. ____________

57. Interconnected food chains are usually formed by ____________ (producers/omnivores).

58. Identify the number of grazing food chains in the diagram of a food web given below.

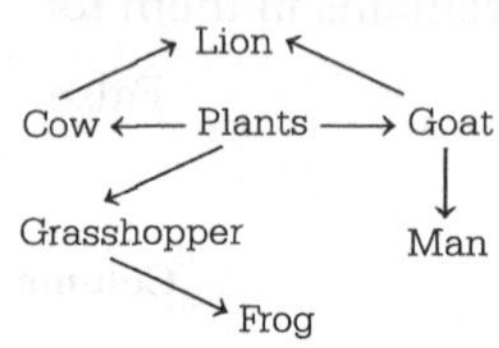

____

59. Give the biological term for the specific place of an organism in the food chain based on their nutrition. ______________________________

60. Refer to the flowchart of a typical food chain and answer the following questions.

Ⓐ Primary producers → Ⓑ Primary consumers → Ⓒ Secondary consumers → Ⓓ Tertiary consumers.

(i) Mention the trophic levels of *A* and *D*.

____

(ii) Give the examples of organisms which are likely to present at first trophic level. ____________

(iii) The organisms found at position *B* are likely to be ____________ (carnivores/herbivores)

(iv) At which position, birds and fishes would be found?

____

(v) Zooplanktons being primary consumers would occupy ____________ trophic level.

61. The amount of energy increases at successive trophic levels.

True ☐ False ☐

62. The mass of living material at a trophic level at a particular time is known as

____

63. Refer to the given statements and identify the biological term.

I. It is the mass of living matter in a unit area.

II. It is expressed in terms of fresh or dry weight.

____

64. Refer to the diagram of energy flow through different trophic levels and answer the following questions.

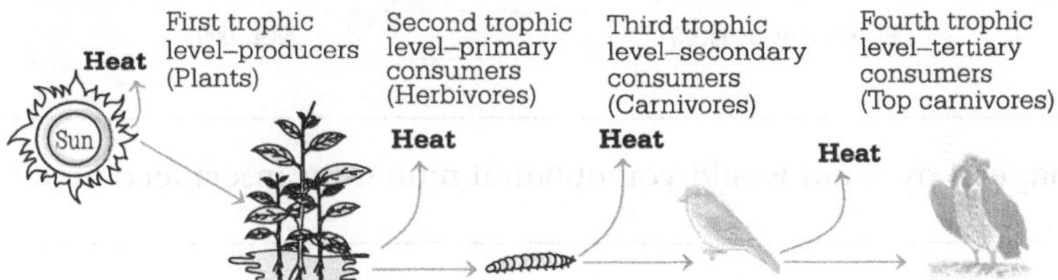

(i) Identify the type of food chain shown in the given diagram.

____

(ii) What per cent of energy is being transferred from first to second trophic level?

____

(iii) If the plants received 5000 kJ energy from sun, what amount of energy would be received by the top carnivore?

____

(iv) The number of trophic levels in a grazing food chain is limited due to 10% law of energy transfer.

True ☐ False ☐

Topic 3 : Ecological Pyramids

65. Give the biological term for the graphical representation of producers and consumers relationship in the form of a pyramid.

____

66. Choose the odd one out with respect to the applicability of ecological pyramids.

Number, Biomass, Energy, Succession

____

67. The base of an ecological pyramid represents top carnivores.
(*Rewrite the statement by changing the underlined word.*)

____

68. While calculating energy, biomass or number, all the organisms at a trophic level of pyramid are considered.

True ☐ False ☐

69. The trophic level in ecological pyramids represents ______________ (a species/ functional level).

70. In most ecosystems, pyramid of number, biomass and energy is ______________ .

71. What does the ecological pyramid shown in the given diagram represents?

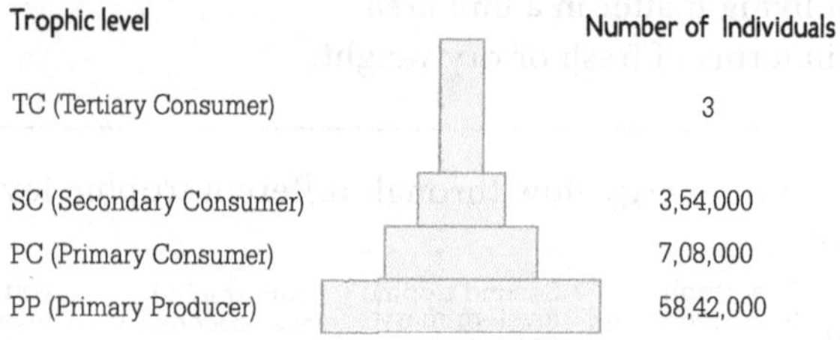

____

72. What shape of pyramid would you obtain if numerous insect feed on a big tree?

____

73. If a tree support number of small birds which inturn are eaten by large birds, then the shape of pyramid of number would be

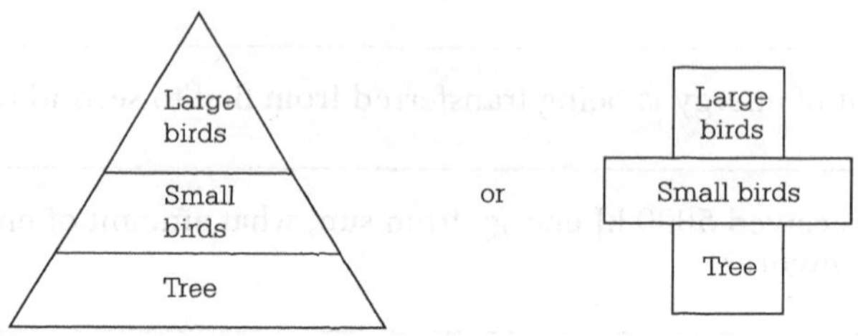

A *B*

____

74. Refer to the given diagram of pyramid of biomass.

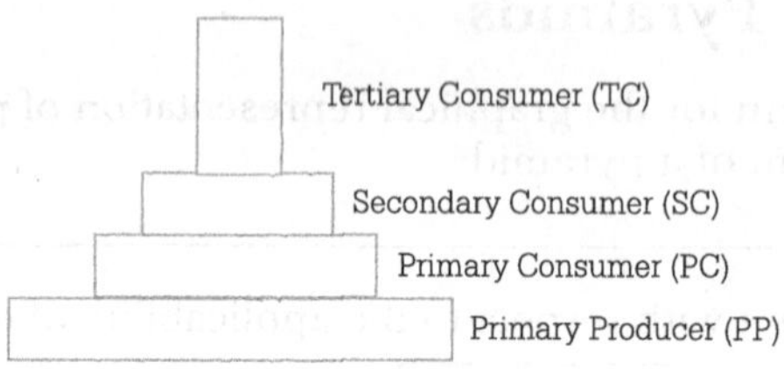

The dry weight (kg m^{-2}) of biomass from PP to TC would ______________ (increase/decrease).

75. What shape of pyramid of biomass would you obtain if small standing crop of phytoplanktons supports large standing crop of zooplanktons?

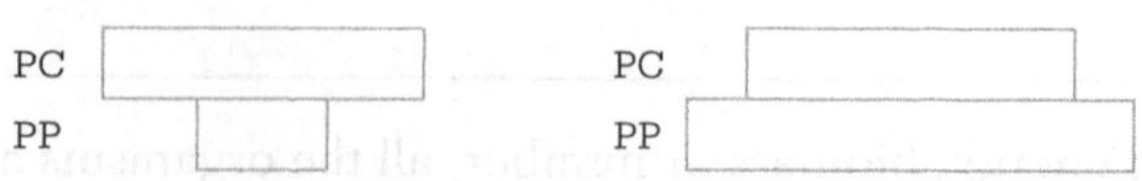

A *B*

____

76. In a sea, the pyramid of biomass is ______________ shaped.

77. An ideal pyramid of energy is

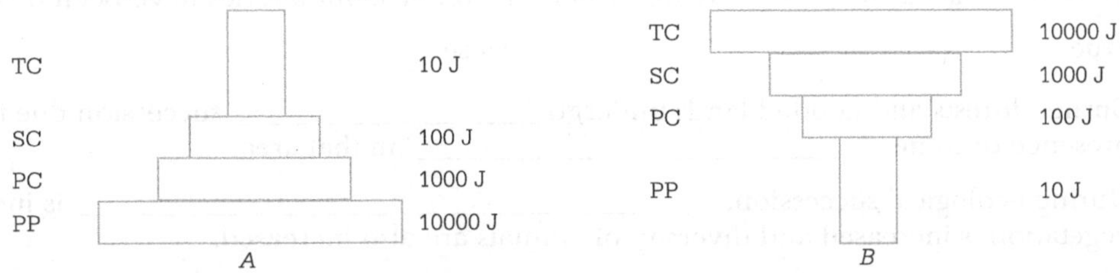

____

78. How many type of pyramids listed below occur in nature?

Inverted pyramid of number, Upright pyramid of biomass, Inverted pyramid of biomass, Inverted pyramid of energy, Spindle-shaped pyramid of number.

____

79. Ecological pyramids take into account food ____________ and does not accommodate a food ____________ .

80. Ecological pyramids neglect the role of ________________ (saprophytes/herbivores) in the ecosystems.

Topic 4 : Ecological Succession

81. Match the following columns.

Column I	Column II
A. Ecological succession	1. Sequence of communities that change successively
B. Sere	2. Stable community
C. Climax community	3. Gradual changes in species composition of given area
D. Pioneer community	4. Unstable community

A. ____________ B. ____________ C. ____________ D. ____________

82. Complete the following analogy given below with respect to succession.
Pioneer community : Little diversity and fast growth : : ____________________ : More diversity and little growth

83. Give the biological term for the individual transitional communities occurring during ecological succession.

____

84. The present day communities in the world are formed due to the parallel processes of succession and ____________________ .

85. The type of succession where no living organism ever existed is known as

____

86. Choose the odd one out with respect to the site of primary succession.

Bare rock, Newly cooled lava, Newly created pond, Abandoned farm land

____

87. The establishment of biotic community is very slow during primary succession due to the absence of ____________________ (fertile soil/autotrophs).

88. Secondary succession occurs in area where the biotic communities have been destroyed.

True ☐ False ☐

89. Burned forests and flooded lands undergo ______________ succession due to the presence of some ______________________ in that area.

90. During ecological succession, ______________________________ is increased, vegetation is increased and diversity of animals are also increased.

91. Complete the following analogy given below with respect to type of succession.

Autogenic succession : Changes brought by organisms themselves : :
____________________: Changes brought by external conditions.

92. Seral stages of succession are irreversible.

True ☐ False ☐

93. Match the following columns.

Column I	Column II
A. Xerarch succession	1. Medium water conditions
B. Mesic condition	2. Occurs in wet areas
C. Hydrarch succession	3. Occurs in dry areas

A. ______________ B. ______________ C. ______________

94. Give the biological term for the species that invade a bare area. ______________

95. Name the pioneer species in primary succession on rocks. ______________

96. How many of the listed functions are performed by lichens during primary succession on rocks?

Acid secretion, Rock formation, Soil formation, Weathering of rocks

____

97. Arrange the following stages of primary succession on rocks in the order of their occurrence.

I. Mesophytic conditions, II. Bryophytes, III. Forest, IV. Lichens

____

98. Complete the following analogy given below.

Lichens : Pioneer species on rocks : ______________________ : Pioneer species in water

99. Refer to the diagram of various stages of primary succession in water and answer the following questions.

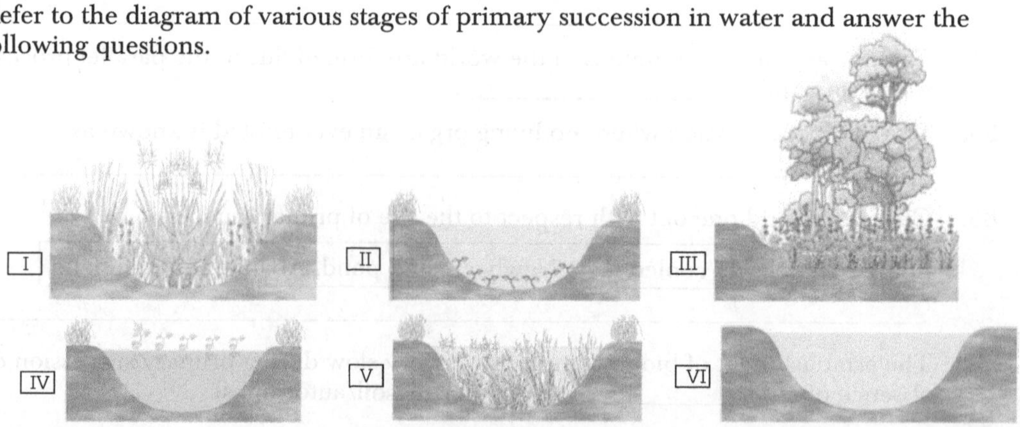

(i) Arrange the following in the order of their occurrence during succession.

____

(ii) What does the stage V called ? _______________

(iii) The stage I is known as _______________ .
(shrub stage/marsh-meadow stage)

(iv) Identify the phytoplankton stage in the given diagrams. _______________

(v) Give an example each of the stage shown in the diagrams.

____

100. The diagram given below represents the _______________ community in xerarch succession.

101. How many of the listed factors affect the secondary succession in an area?

Soil condition, Water availability, Nutrient cycle, Environment, Propagules

____

102. The climax community of hydrarch and xerarch successions possesses _______________ conditions.

103. Match the following columns.

Column I (Terms related to succession)	Column II (Descriptions)
A. Nudation	1. Successful establishment of a species in a bare area.
B. Invasion	2. Arrival of reproductive propagules in a bare area.
C. Ecesis	3. Individual like affected by surround species.
D. Co-action	4. Development of bare area without any form of life.

A _______________ B. _______________ C. _______________ D._______________

104. Choose the odd one out with respect to the species found in lithosere.

Rinodina, Parmelia, Polytrichum, Vallisneria

____

Topic 5 : Nutrient Cycle and Ecosystem Services

105. The amount of nutrients, e.g. carbon, phosphorus, etc., present in the soil at any given time is referred to as _______________ (biomass/standing state).

106. Give the biological term for the movement of nutrient elements through various components of an ecosystem.

____

107. Complete the following analogy given below with respect to biogeochemical cycle.
Gaseous nutrient cycle : Reservoir in atmosphere : : Sedimentary nutrient cycle :

____

108. Sulphur and phosphorus cycles are the gaseous type of nutrient cycle.
(*Rewrite the statement by changing underlined word.*)

____

109. Environmental factors, e.g. soil, pH, temperature, etc., regulate the release of __________ into the atmosphere. The function of __________ is to meet the deficit which occur due to __________ in the rate of influx and efflux of nutrients.

110. Complete the following analogy given below with respect to carbon fixation by photosynthesis.
Amount of carbon in organisms : 49% : : Amount of carbon dissolved in oceans : ________

111. What amount of total global carbon is found in atmosphere?__________

112. Refer to the simplified model of carbon cycle in the biosphere and answer the following questions.

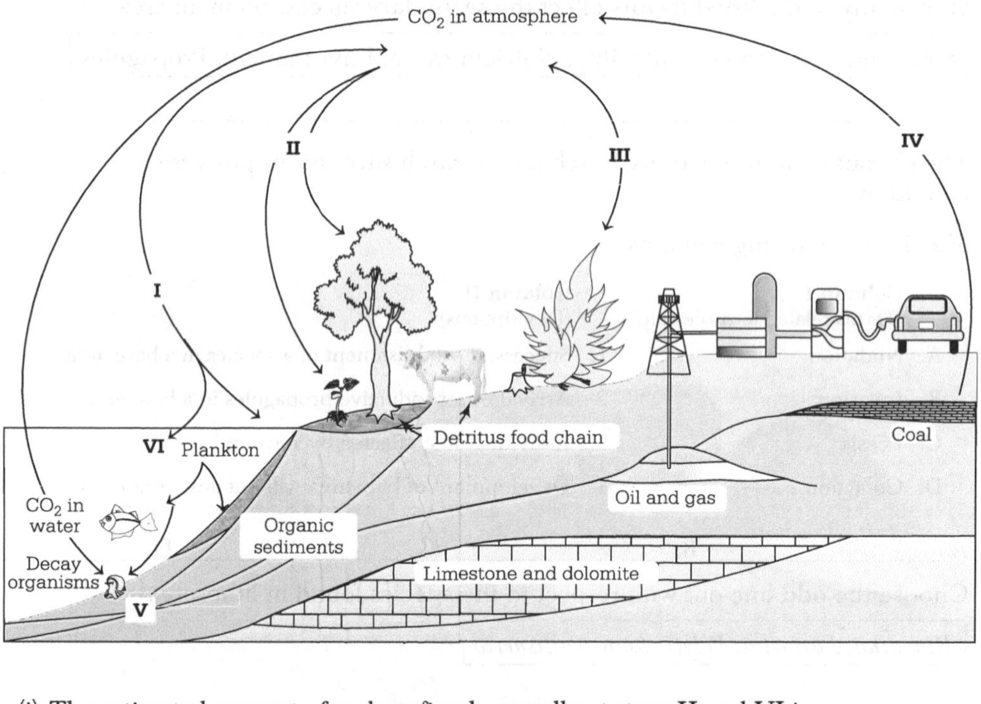

(i) The estimated amount of carbon fixed annually at steps II and VI is __________. (4×10^{13} kg / 45×10^{15} kg)

(ii) At which step, the carbon is removed from the circulation? __________

(iii) Carbon is released into atmosphere at steps I, III and IV as carbon __________ (monoxide/dioxide).

113. Name any two reservoirs of carbon.

____

114. Decomposers help to ________________ (fix/release) CO_2 by processing waste material and dead organic matter.

115. Human activities like deforestation increase the rate of carbon fixation.

True ☐ False ☐

116. Mention any two functions of phosphorus in animals.

(i) ________________ (ii) ________________

117. The natural reservoir of phosphorus is <u>ocean</u>.
(*Rewrite the statement by changing the underlined word.*)

____

118. The phosphorus cycle circulates phosphorus in between ________________ components .

119. Refer to the simplified model of phosphorus cycling and answer the following questions.

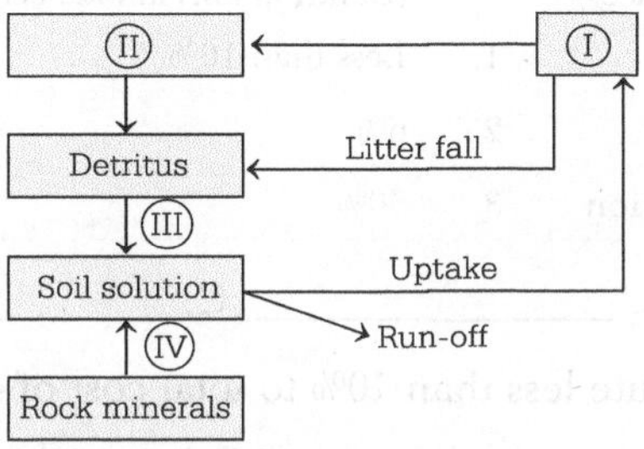

(i) The given cycle is likely to be found in ________________ ecosystem.

(ii) Identify the organism labelled as I and II. ________________

(iii) Name the processes labelled as III and IV. ________________

120. Match the following columns.

Column I	Column II
A. Herbivores	1. Release phosphorus from dead and decaying material.
B. Phosphate solubilising bacteria	2. Obtain phosphorus from plants.
C. Producers	3. Likely to contain maximum amount of phosphorus.

A. ____________ B. ____________ C. ____________

121. Name the process which releases major amount of phosphorus into atmosphere.

____

122. Unlike ________________ cycle, there is no respiratory release of ________________ into atmosphere.

123. Atmospheric inputs of phosphorus through rainfall are much smaller than carbon inputs.

True ☐ False ☐

124. Give the biological/ecological term for the products of ecosystem processes.

____

125. How many of the listed term below are ecosystem services?

Fertile soil, Crop pollination, Ecological pyramids, Carbon storage, Wildlife habitat

____

126. Who attempted to put price tags on nature's life support services?

____

127. An average price tag on fundamental ecosystem services per year is about _______________ (US $33 trillion/US $ 3 million).

128. The cost of fundamental ecosystem services is lesser than the value of global gross national product.

True ☐ False ☐

129. Match the following columns.

	Column I (Ecosystem services)		Column II (Contributions in total ecosystem service case)
A.	Soil formation	1.	Less than 10%
B.	Recreation	2.	6%
C.	Climate regulation	3.	50%

A. _______________ B. _______________ C. _______________

130. Nutrient cycling contribute less than 10% to total cost of ecosystem services.

True ☐ False ☐

131. The cost of wildlife habitat is about _______________ (6%/60%) of the total cost of an ecosystem services.

Answers

1. Functional, physical environment, AG Tansley.
2. Terrestrial and aquatic ecosystems.
3. Estuary is odd one out as it is an aquatic ecosystem whereas the rest three are terrestrial ecosystems.
4. Crop fields and aquarium.
5. Biotic and abiotic components.
6. Stratification
7. Top, grasses.
8. 4 (Productivity, decomposition, energy flow, nutrient cycling). Stratification and biome are not function of ecosystem.
9. True
10. *A*- Abiotic components
    B- Producers
    C- Zooplanktons
    D- Fungi, bacteria and flagellates
11. (i) Water
    (ii) Autotrophs
    (iii) Decomposers
12. Decomposition and mineralisation
13. Unidirectional, higher, heat.
14. A-3, B-1, C-4, D-2
15. $gm^{-2}\ yr^{-1}$
16. Net Primary Productivity (NPP).
17. 4 (Plant species habitat, environmental factors, nutrient availability, photosynthetic capacity of plants).
18. 55 billion tons
19. False. Primary productivity of different ecosystem varies and the oceans has low productivity despite occupying 70% of the surface.
20. Decomposition
21. Detritus
22. Inorganic substances is odd one out as it is the end product of decomposition.
23. Detritivores
24. A-3, B-4, C-1, D-2, E-5
25. Earthworm
26. (i) V $\rightarrow$ I $\rightarrow$ II $\rightarrow$ III $\rightarrow$ IV
    (ii) 3 (I, II, V)
    (iii) 2 (III, IV)
27. False. Humus is highly resistant to microbial action and undergoes decomposition at slower rate.
28. Humus
29. Decomposition cycle in terrestrial ecosystem.
30. Oxygen, chemical composition, climatic
31. Nitrogen and sugars in detritus
32. Temperature and soil moisture
33. Anaerobiosis is the odd one out as it inhibits decomposition whereas the other two factors favour decomposition.
34. True
35. Inhibits
36. Deep sea hydrothermal ecosystem
37. Photosynthetically Active Radiation
38. 2-10%
39. Unidirectional
40. False. Ecosystems are not exempt from the second law of thermodynamics, they need a constant supply of energy to synthesise the molecules.
41. Producers
42. Food chain
43. Producers
44. False. Energy trapped by plants passed on to consumers and thus, it does not remain in them forever.
45. A-3, B-1, C-2
46. 3 (Insects, birds, mammals). Molluscs are the herbivores in aquatic ecosystem.
47. Tertiary consumers
48. Grazing Food Chain
49. (i), i.e. Grass $\rightarrow$ Goat $\rightarrow$ Man
50. Producers
51. Detritus Food Chain
52. Organic matter
53. Fungi and bacteria
54. Saprotrophs
55. Grazing, detritus
56. Food webs
57. Omnivores
58. 4 (i) Plants $\rightarrow$ Cow $\rightarrow$ Lion
    (ii) Plants $\rightarrow$ Goat $\rightarrow$ Lion
    (iii) Plants $\rightarrow$ Grasshopper $\rightarrow$ Frog
    (iv) Plants $\rightarrow$ Goat $\rightarrow$ Man
59. Trophic level
60. (i) *A*–First trophic level, *D*–Fourth trophic level
    (ii) Phytoplanktons, grass, trees
    (iii) Herbivores
    (iv) *C*, i.e. secondary consumers
    (v) Second
61. False. The amount of energy decreases at successive trophic levels.
62. Standing crop
63. Biomass
64. (i) Grazing food chain
    (ii) 10%. According to Lindemann, only 10% of available energy at a trophic level passes to the next level. The rest of energy is either consumed or dissipated.
    (iii) 5 kJ. In the given diagram, top carnivore is present at fourth trophic level and each level receives only 10% energy from the lower trophic level.

    Plants 5000 kJ $\rightarrow$ Herbivores 500 kJ $\rightarrow$ Carnivores 50 kJ $\rightarrow$ Top carnivores 5 kJ

    (iv) True
65. Ecological pyramids
66. Succession is odd one out as it is not studied through ecological pyramids.
67. Producers

68. True

69. Functional level

70. Upright

71. Pyramid of number in grassland ecosystem.

72. Inverted pyramid of number.

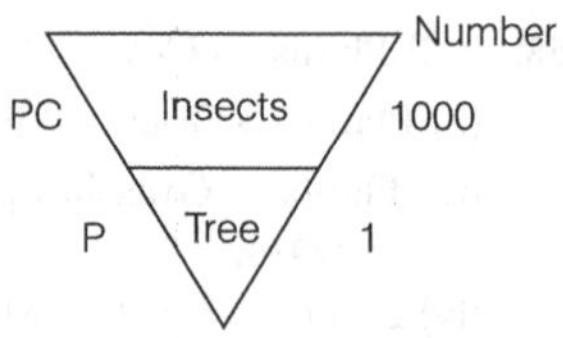

73. *B* Spindle shape, because number of tree would be less than smaller birds. Also the population of large birds would be lesser than that of smaller birds.

74. Decrease

75. *A*, i.e. inverted pyramid because standing crop of phytoplanktons is lesser than that of zooplanktons.

76. Inverted

77. *A*-Pyramid of energy is always upright.

78. 4 (Inverted pyramid of number, upright pyramid of biomass, inverted pyramid of biomass, spindle-shaped pyramid of number). Inverted pyramid of energy is never found because some energy is always lost as heat at each trophic level.

79. Chain, web

80. Saprophytes

81. A-3, B-1, C-2, D-4

82. Climax community

83. Seral stages or seral communities.

84. Evolution

85. Primary succession

86. Abandoned farm land is odd one out as it is an example of secondary succession.

87. Fertile soil

88. True

89. Secondary, soil or sediments

90. Niche specialisation

91. Allogenic succession

92. False. Any disturbance (natural or induced) can convert a particular seral stage to an earlier stage and thus, it is reversible.

93. A-3, B-1, C-2

94. Pioneer species

95. Lichens

96. 3 (Acid secretion, soil formation, weathering of rocks). Lichens do not help in rock formation, rather they help to dissolve the rocks.

97. IV $\rightarrow$ II $\rightarrow$ III $\rightarrow$ I.

98. Phytoplanktons

99. (i) VI $\rightarrow$ II $\rightarrow$ IV $\rightarrow$ V $\rightarrow$ I $\rightarrow$ III

(ii) Reed-swamp stage

(iii) Marsh-meadow stage

(iv) VI

(v) VI - Plankton stage
- Diatoms, Green flagellates.

II - Submerged stage
- *Hydrilla, Vallisneria.*

IV - Submerged free-floating stage - *Nymphaea, Wolffia.*

V - Reed swamp stage
- *Typha, Sagittaria.*

I - Marsh - meadow stage
- *Polygonum, Dichanthium.*

III - Scrub stage
- *Cephalanthus.*

100. Climax

101. 4 (All of them except nutrient cycle).

102. Mesic

103. A-4, B-2, C-1, D-3

104. *Vallisneria* is odd one out, as it is found in hydrarch succession during submerged stage.

105. Standing state

106. Nutrient cycle or biogeochemical cycle.

107. Reservoir in earth's crust

108. Sedimentary

109. Nutrients, reservoir, imbalance

110. 71%

111. 1%

112. (i) 4×10^{13} kg

(ii) V. Some amount of the fixed carbon is lost to sediments and removed from circulation.

(iii) Dioxide

113. Oceans and fossil fuels

114. Fix

115. False. Human activities like deforestation increases the rate of release of carbon into the atmosphere.

116. (i) To make shells, bones and teeth.

(ii) Constituent of biological membranes and nucleic acids.

117. Rock

118. Biotic and abiotic.

119. (i) Terrestrial

(ii) I-Producers, II-Consumers

(iii) III-Decomposition, IV-Weathering

120. A-2, B-1, C-3

121. Weathering

122. Carbon, phosphorus

123. True

124. Ecological services

125. 4 (All of them except ecological pyramids). Ecological pyramid is not an ecosystem services, rather it represents ecological structure.

126. Robert Constanza

127. US $ 33 trillion

128. False. The value of ecosystem services is twice the value of the global gross national product, which is US $ 18 trillion.

129. A-3, B-1, C-2

130. True

131. 6%

CHAPTER FIFTEEN

BIODIVERSITY AND CONSERVATION

Topic 1 : Biodiversity : Levels and Patterns

1. The term 'Biodiversity' was coined by ________________ (Mayer/Walter G Rosen).
2. Name the sociobiologist who popularised the term biodiversity. ____________
3. Biodiversity refers to the combined diversity at three major levels *viz.* ____________ (genetic, species, ecological/atomic, molecular, genetic)
4. Heterogeneity refers to the immense ________________ in biosphere, at all the levels of biological organisation.
5. The significance of biodiversity existing on earth can be understood by the efforts made by ________________ and ______________ biologists.
6. On earth, there are exist more than 20,000 species of ___________ and ___________ .
7. A single species cannot show high diversity at the genetic level over its distributional range.

   True ☐ False ☐
8. The medicinal plant, ________________ growing in Himalayan ranges shows genetic diversity in terms of potency and ________________________ of the active chemical, i.e. ______________________ produce by it.
9. India has more than <u>2000</u> genetically different strains of rice.
   (*Rewrite the statement by changing the underlined word.*)

   ____
10. More than 2000 genetically different varieties of mango are found in India.

    True ☐ False ☐
11. More amphibian species in Western ghats than the Eastern ghats is an example of

    ____
12. Ecological diversity increases with the type of ______________ diversity.

13. Scandinavian country like Norway has less coral reefs, estuaries and alpine meadows as compared to India.

True ☐ False ☐

14. Choose the odd one out with respect to the type of diversity at ecosystem level.

Alpha (α) diversity, Beta (β) diversity, Gamma (γ) diversity, Sigma (σ) diversity

____

15. Match the following columns.

Column I (Levels of biodiversity)	Column II (Characteristics of levels)
A. α-diversity	1. Diversity in between the two communities
B. β-diversity	2. Richness of species in all habitats
C. γ-diversity	3. Diversity within community

A.________ B.________ C. ________

16. Name the scientist who proposed the concept of α, β and γ-diversities.

____

17. Complete the analogy given below with respect to biodiversity.

α-diversity : Local diversity : : ________ : Regional diversity

18. Expand IUCN.

____

19. According to IUCN, the total number of plant and animal species on earth is about ________ (1.5 million/15 billion).

20. Species inventories for many taxonomic groups are more complete in <u>tropical</u> countries. (*Rewrite the statement by changing the underlined word.*)

____

21. According to Robert May, the global species diversity on earth is about ________ . The latest study conducted in 2012, estimated about ________ terrestrial and ________ species in oceans.

22. Based on current species inventories, more than 70% of all species recorded are plants, while animals comprise about 22% of the total.

True ☐ False ☐

23. Refer to the diagram showing global biodiversity of invertebrates and answer the questions that follows.

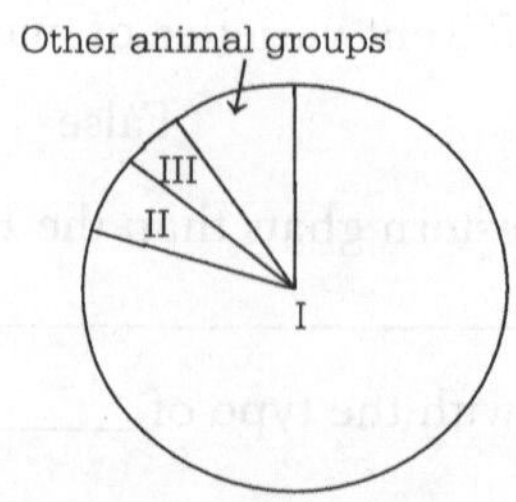

(i) Match the following columns with respect to above figure.

	Column I (Groups of animal labels)		Column II (Examples of groups)
A.	I	1.	Molluscs
B.	II	2.	Insects
C.	III	3.	Crustaceans

A. __________ B. __________ C. __________

(ii) According to the given pie chart, most species-rich taxonomic group is I, which make __________ (70/90) per cent of the total diversity on earth.

24. (i) Refer to the diagram of vertebrates biological biodiversity and identify the taxonomic groups labelled as *A* and *B*.

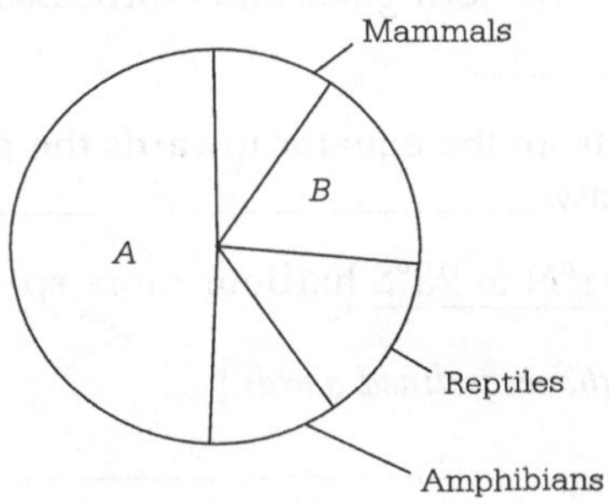

A. __________

B. __________

(ii) The number of species of taxonomic group '*A*' is more than the combined total of the species of birds, amphibians, reptiles and mammals.

True ☐ False ☐

25. Given below diagram represents global biodiversity of plants on earth and answer the questions that follows.

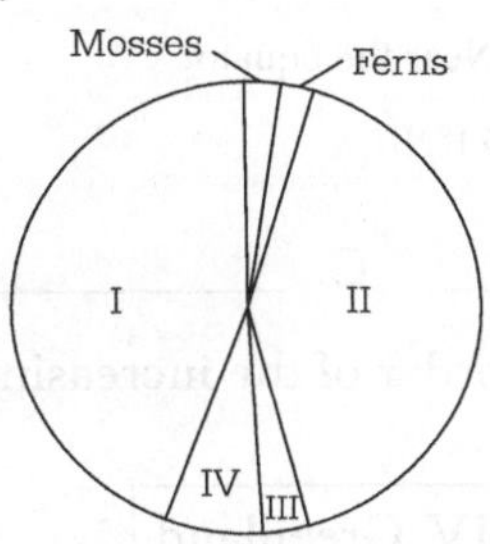

(i) Match the following columns with respect to figure shown above.

	Column I (Plant groups lebels)		Column II (Examples of plant group)
A.	I	1.	Lichens
B.	II	2.	Fungi
C.	III	3.	Algae
D.	IV	4.	Angiosperms

A. __________ B. __________ C. __________ D. __________

(ii) The given diagram does not give any figures for __________ (gymnosperms/prokaryotes)

26. The conventional taxonomic methods are not suitable to determine the global biodiversity of ________________ species.

27. What share does India contribute to global species diversity, if its world's land area is 2.4%?

____

28. The number of megadiversity countries in the world is ________________

29. According to May's global estimates, what per cent of total species have been recorded so far?

____

30. Complete the following analogy with respect to the regions of biodiversity.

Biodiversity rich region in India : Western ghats and North-East :: Largest biogeographical region in India : ________________

31. 'The species diversity decreases from the equator towards the poles'. Give the ecological term for this pattern of biodiversity. ________________

32. The tropics latitudinal range of 71°N to 23°S harbour more species than temprate or polar areas.
(*Rewrite the statement by changing the underlined words.*)

____

33. Match the following columns.

Column I (Places)	Column II (Latitudinal gradients)
A. Colombia	1. 41°N
B. New York	2. Tropical latitudes
C. Greenland	3. Near the equator
D. India	4. 71°N

A. ________ B. ________ C. ________ D. ________

34. Arrange the following countries in the order of the increasing number of bird species found in them.

I. India, II. Colombia, III. New York, IV. Greenland

____

35. Complete the following analogy with respect to regionwise biodiversity.

Forest in tropical region : ________________ :: Forest in temperate region : Midwest of the USA.

36. Equador forest has less number of vascular plant species as compared to the midwest forest of the USA.

True ☐ False ☐

37. Name the tropical forest in South America which has greatest biodiversity on earth.

____

38. Match the following columns.

	Column I (Species)		Column II (Numbers in Amazonian rainforest)
A.	Plants	1.	3000
B.	Fishes	2.	378
C.	Mammals	3.	40,000
D.	Birds	4.	427
E.	Reptiles	5.	1300

A.________ B.________ C. ________ D. ________

E.________

39. The number of species of invertebrates in Amazonian rainforest is greater than the combined number of plants and vertebrates.

True ☐ False ☐

40. How many of the listed factors are found in tropical regions and contribute to their greater biological diversity?

Frequent glaciations, Less seasonal and more constant, More solar energy, Niche specialisation, Unpredictable environment, Higher productivity

____

41. Species-area relationship was proposed by the geographer ________ during his exploration in the jungles of ________

42. Refer to the graph showing species-area relationship and answer the questions that follows.

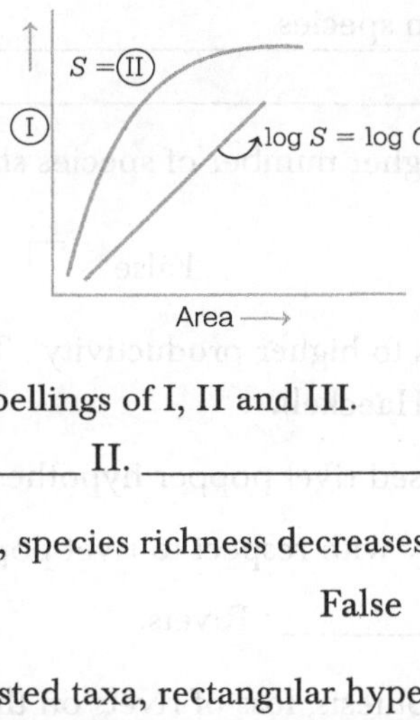

(i) Mention the correct labellings of I, II and III.

I. ________ II. ________ III. ________

(ii) As per the given graph, species richness decreases with increasing explored area.

True ☐ False ☐

(iii) For how many of the listed taxa, rectangular hyperbola is obtained?

Angiosperm plants, Bryophytes, Birds, Bats, Freshwater fishes, Porifera

____

43. Study the following equation with respect to species-area relationship graph which is become linear on log scale and represented as

$\log S = \log C + Z \log A$

Identify *S*, *C* and *Z* in the equation.

____

44. Choose the odd one out with respect to the *Z*-value of species-area relationship.

Plants in Britain, Birds in California, Frugivorous birds in tropical forest, Molluscs in New York

____

45. The *Z*-value for the mammals in the tropical forest lies with a steeper line of <u>2</u>. (*Rewrite the statement by changing the underlined word.*)

____

46. Find the value of log '*S*', if log $A = 4$, $Z = 0.3$ and log $C = 0.8$.

____

47. The steeper slope in species-area relationship appear when very large areas are considered.

True ☐ False ☐

Topic 2 : Importance and Loss of Biodiversity

48. Ecologist believed communities with more species-tend to be more ______________ (stable/unstable) than those with less species.

49. Choose the odd one out with respect to the characteristics of a stable community.

Variable productivity, Resistance to natural and man-made disturbances, Resistance to invasion by alien species

____

50. According to David Tilman, higher number of species show higher year to year variation in total biomass.

True ☐ False ☐

51. 'Increased diversity contributes to higher productivity'. This statement was proposed by ____________ (David Tilman/Haeckel).

52. Name the ecologist who proposed rivet popper hypothesis. ______________________

53. Complete the following analogy with respect to rivet popper hypothesis.

Ecosystem : Airplane :: ______________ : Rivets.

54. According to rivet popper hypothesis, loss of rivets on the wings of airplane is comparable to the loss of ___________________________ in ecosystem.

55. Rivet popper hypothesis explains that the loss of species affect the proper functioning of ecosystem.

Yes ☐ No ☐

56. As per rivet popper hypothesis, loss of few rivets from which part (I or II) would not pose serious threat to ecosystem, i.e. aeroplane shown below.

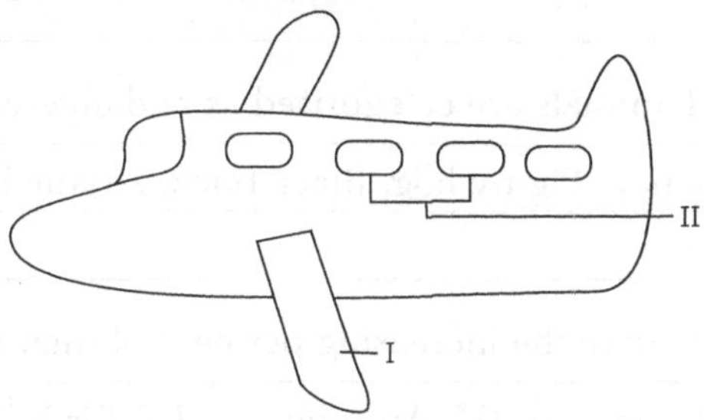

____

57. The colonisation of tropical Pacific Island by humans have led to the extinction of ____________ species of native ____________ .

58. Name the list prepared by IUCN to document the taxa facing risk of extinction.

____

59. How many species of vertebrates, invertebrates and plants have been extinct according to Red list 2004?

____

60. Match the following columns.

Column I (Recent extinctions)	Column II (Countries)
A. Dodo	1. Australia
B. Quagga	2. Mauritius
C. Thylacine	3. Russia
D. Stellar's sea cow	4. Africa

A.__________ B.__________ C. __________ D. __________

61. Bali, Javan and Caspian are three extinct subspecies of Rhinoceros.
(*Rewrite the statement by changing the underlined word.*)

____

62. Name the group of animal which is more vulnerable to extinction.

____

63. When and where the recent Red list was released?

____

64. Complete the following analogy given below.
Extinct in the wild : Survive in cultivation only : : ____________ : High risk of extinction in the wild in future

65. Choose the odd one out with respect to the category of listed plants in Red list.

Berberis nilghiriensis, Nepenthes, Cycas beddomei, Podophyllum

____

66. How many of the listed animals are categorised as endangered in the recent Red list ?

Siberian crane, Red panda, Pigmy hog, Black buck, Asiatic lion

____

67. Arrange the following taxa in the increasing per cent of their threat to extinction.

I. Gymnosperms, II. Mammals, III. Amphibians, IV. Birds

____

68. How many episodes of mass extinction had occurred since the origin and diversification of life on earth? ____________

69. The present extinction episode is considered as the ________ extinction. It is estimated to be 100 to 1000 times ________ than in the pre human times due to ________ activities.

70. Loss of biodiversity leads to ____________ in plant production ____________ resistance to environmental perturbations and ____________ variability in plant productivity, water use, disease cycle, etc.

____

71. Give the sorbiquet used to describe the four major causes of biodiversity losses.

____

72. The most important cause of animals and plants being driven to extinction is

____

73. Match the following columns.

	Column I (Causes of biodiversity loss)		Column II (Examples)
A.	Habitat loss and fragmentation	1.	Overharvesting of marine fishes
B.	Overexploitation	2.	Extinction of Cichlid fish due to Nile perch in Lake Victoria
C.	Alien species invasion	3.	Decreased per cent of land surface covered by tropical forest (14%-6%)
D.	Co-extinction	4.	Coevolution of plant and pollinator

A. ____________ B. ____________ C. ____________ D. ____________

74. The Amazon rainforest is being cut and cleared for cultivating ____________ or for conversion to ____________ for raising ____________ . It leads to the degradation of large habitats into ____________ .

75. How many of the following species have been extinct due to their overexploitation by humans?

Stellar's sea cow, Nile perch, Passenger pigeon, *Lantana*

____

76. Give the common term for the following species.

Parthenium, Lantana, Eichhornia

____

77. Complete the following analogy given below with respect to alien species invasion.

Nile perch : Cichlid fish : : ____________ : Indigenous catfish

78. Co-extinction results due to the facultative relationship between species.

True ☐ False ☐

Topic 3 : Biodiversity Conservation

79. Narrowly utilitarian, broadly utilitarian and ethical argument are the three categories of biodiversity ____________ (pattern/conservation).

80. Match the following columns.

	Column I (Causes of biodiversity conservation)		Column II (Importances)
A.	Narrowly utilitarian	1.	Ecosystem services provided by nature
B.	Broadly utilitarian	2.	Every species has an intrinsic value
C.	Ethical argument	3.	Direct economic benefits from nature

A.____________ B.____________ C. ____________

81. Choose the odd one out with respect to the type of biodiversity conservation categories.

Drugs, Food, Pollination, Medicinal plants

____

82. Give biological term for exploring molecular, genetic and species level diversity for economic important products.

____

83. 'The aesthetic pleasure of walking through thick woods is an example of <u>ethical argument</u>'.

(*Rewrite the statement by changing the underlined word*.)

____

84. Ethical argument of biodiversity conservation apprise to conserve every species even it has no economic value to us.

True ☐ False ☐

85. Complete the following analogy given below.
On-site conservation : ____________ : : Off-site conservation : *Ex situ*

86. Which approach of conservation is employed when a species is endangered and need urgent measures to save it from extinction?

____

87. Give the biological term for the species confined to particular region and not found elsewhere.

____

88. Choose the odd one out with respect to the characteristics of hotspots.

Low species richness, High endemism, Accelerated habitat loss, High biodiversity

____

89. The total number of hotspots in world is ____________

90. How many hotspots are found in India? ____________

91. All the listed areas are hotspots in India except one, that is

Western Ghats and Sri Lanka, Indo Burma and Himalaya, Madagascar

____

92. All biodiversity hotspots collectively cover less than ____________ per cent of earth's land area. They harbour ____________ number of species. The strict protection of hotspots can reduce ongoing ____________ by almost 30%.

93. What type of conservation strategy is employed in biosphere reserves and national parks?

____

94. Arrange the following conservation sites in the increasing order of their number in India.

I. National parks, II. Wildlife sanctuaries, III. Biosphere reserves

____

95. Match the following columns.

Column I (National Parks of India)	Column II (States)
A. Bandipur National Park	1. Maharashtra
B. Chandoli National Park	2. Kerala
C. Dudhwa National Park	3. Madhya Pradesh
D. Kanha National Park	4. Uttar Pradesh
E. Silent Valley National Park	5. Karnataka

A.____________ B.____________ C. ____________

D.____________ E.____________

96. Complete the following analogy given below.
Kaziranga National Park : Indian Rhinoceros Park :: Dachigam National Park :

____

97. How many of the listed national parks are found in Uttarakhand?

Rajaji National Park, Corbett National Park, Indravati National Park, Nanda Devi National Park

____

98. Specified area zonation, *viz.* core, buffer and manipulation is seen in ______________ (biosphere reserves/national parks).

99. Name the zone of biosphere reserve where
(i) no human activity is allowed. ______________
(ii) limited human activities are allowed. ______________

100. Nanda Devi, Sunderbans, Nilgiri and Gulf of Mannar are biosphere reserves of India which are now recognised as world heritage sites.
True ☐ False ☐

101. Match the following columns.

Column I (Sanctuaries in India)	Column II (Protected species)
A. Keoladeo Bird Sanctuary	1. Olive ridley turtles
B. Gahirmatha Marine Sanctuary	2. One-horned rhino
C. Manas Wildlife Sanctuary	3. Migratory Siberian crane
D. Mudumalai Wildlife Sanctuary	4. Gaur and Indian leopard

A.__________ B.__________ C. __________ D. __________

102. Biodiversity rich and productive Ramsar sites are______________ (low mountain ranges/wetlands).

103. Give biological term for the protected forest patches having religious and cultural significance.______________

104. Sacred groves help in *ex situ* conservation of wildlife and threatened plants.
True ☐ False ☐

105. Match the following columns.

Column I (Sacred groves)	Column II (Locations)
A. Khasi and Jaintia	1. Rajasthan
B. Aravalli hills	2. Karnataka and Maharashtra
C. Western Ghats	3. Madhya Pradesh
D. Sarguja, Chanda and Bastar areas	4. Meghalaya

A.__________ B.__________ C. __________ D. __________

106. In *ex situ* conservation, threatened species are protected in their natural habitats.

True ☐ False ☐

107. Give any three examples of *ex situ* conservation.

____

108. Animals extinct in wild are continued to be maintained in national parks.
(*Rewrite the statement by changing the underlined word*.)

____

109. Rhino conservation project was started in ________________ at ____________________ .

110. Name the technique used to preserve gametes of threatened species in viable condition for long periods.

____

111. For *ex situ* conservation of gametes, eggs are fertilised __________ and plants are propagated using ________________ methods. Seeds of commercially important plants are kept in ________________ .

112. National Bureau of Plant Genetic Resources is a seed bank in ________________ (New Delhi/Gujarat.)

____

113. When 'The Earth Summit' was held to take measures for biodiversity conservation?

____

114. The Earth Summit was held at ________________ which was the historic convention on biological diversity.

115. State the measures to be taken by all nations after Earth Summits.

____

116. The world summit on sustainable development ________________ take pledged to reduce the current loss of biodiversity at ____________, ____________ and ________________ levels.

Answers

1. Walter G Rosen
2. Edward Wilson
3. Genetic, species, ecological
4. Diversity
5. Ecologist, evolutionary
6. Ants, orchids
7. False. A single species can show high diversity at the genetic level over its distributional range.
8. *Rauwolfia vomitoria*, concentration, reserpine
9. 50,000
10. False. More than 1000 genetically different varieties of mango are found in India.
11. Species diversity
12. Ecosystem
13. True
14. Sigma (σ) diversity is odd one out. The other three represent diversity at the level of community and ecosystem.
15. A-3, B-1, C-2
16. Whittaker in 1972.
17. γ-diversity
18. International Union for Conservation of Nature and Natural Resources.
19. 1.5 million
20. Temperate
21. 7 million, 8.7 million, 2.2 million
22. False. Based on current species inventories, about 70% of total recorded species are animals while 22% are of plants.
23. (i) A-2, B-1, C-3
    (ii) 70
24. (i) *A*–Fishes, *B*–Birds
    (ii) True
25. (i) A-2, B-4, C-1, D-3
    (ii) Prokaryotes
26. Microbial
27. 8.1%
28. 12
29. 22%
30. The Deccan Peninsula
31. Latitudinal gradient
32. 23.5° N to 23.5°S
33. A-3, B-1, C-4, D-2
34. IV $\rightarrow$ III $\rightarrow$ I $\rightarrow$ II
    [56] [105] [1200] [1400]
35. Equador
36. False. Equador forest has up to 10 times more species of vascular plants as compared to midwest forest of the USA.
37. Amazonian rainforest
38. A-3, B-1, C-4, D-5, E-2
39. True
40. 4 (Less seasonal and more constant, more solar energy, niche specialisation, higher productivity). Tropical environment is predictable and constant. Frequent glaciation occurred in temperate environment, which led to less biological diversity in that areas.
41. Alexander von Humboldt, South America
42. (i) I - Species richness
    II - CA^z
    III - $Z \log A$
    (ii) False. Species richness increases with increasing explored area within a region, but up to a limit.
    (iii) 4 (Angiosperm plants, birds, bats, freshwater fishes).
43. *S*–Species richness
    C–Y-intercept
    Z–Regression coefficient
44. Frugivorous birds in tropical forest is odd one out because its *Z*-value range is 0.6-1.2.
    For other listed taxa, *Z*-value range is 0.1 to 0.2.
45. 1.15
46. 2.0, $\log S = 0.8 + 0.3 \times 4$
    $= 0.8 + 1.2 = 2.0$
47. True
48. Stable
49. Variable productivity is odd one out, because a stable community does not show too much variations in productivity from year to year.
50. False. David Tilman proposed that higher number of species show less year to year variation in total biomass.
51. David Tilman
52. Paul Ehrlich
53. Species
54. Key species
55. Yes
56. II, As loss of rivets of wings-I (key species) is of major concern because they drive aeroplane signifying here major ecosystem functions.
57. 2000, birds
58. Red list
59. Vertebrates-338, Invertebrates-359, Plants-87
60. A-2, B-4, C-1, D-3
61. Tiger
62. Amphibians
63. 19 July, 2012 at Rio Earth Summit
64. Endangered
65. *Cycas beddomei* is odd one out, because it is critically endangered plant whereas the other three are endangered plants in Red list.
66. 3 (Red panda, Pigmy hog, Asiatic lion).
    Siberian crane is critically endangered and black busk is threatened species.
67. IV $\rightarrow$ II $\rightarrow$ I $\rightarrow$ III
    (12%) (23%) (31%) (32%)

68. Five

69. Sixth, faster, human

70. Decline, lowered, increased.

71. The Evil Quartet

72. Habitat loss and fragmentation

73. A-3, B-1, C-2, D-4

74. Soybeans, grasslands, beef cattles, small fragments

75. 2 (Stellar's sea cow and passenger pigeon). Rest two are invasive species.

76. Invasive weed species

77. African catfish or *Clarias gariepinus.*

78. False. Co-extinction results due to the obligatory relationship between species.

79. Conservation

80. A-3, B-1, C-2

81. Pollination is odd one out as it is an example of broadly utilitarian, while other three are the example of narrow utilitarian.

82. Bioprospecting

83. Broadly utilitarian.

84. True

85. *In situ*

86. *Ex situ* conservation

87. Endemic species

88. Low species richness is an odd one out because hotspots are characterised by high species richness.

89. 34

90. Three

91. Madagascar is a hotspot, but not in India.

92. Two, high, mass extinction

93. *In situ* conservation

94. III → I → II
[14 (now 18)] [90] [448]

95. A-5, B-1, C-4, D-3, E-2

96. Kashmiri stag

97. 3 (All of them except Indravati National park). Indravati National Park is located in Chhattisgarh.

98. Biosphere reserves.

99. (i) Core zone, (ii) Buffer zone

100. True

101. A-3, B-1, C-2, D-4

102. Wetlands

103. Sacred groves

104. False. Sacred groves help in *in situ* conservation of wildlife and threatened plants.

105. A-4, B-1, C-2, D-3

106. False. In *ex situ* conservation, threatened species are protected outside their natural habitat in special setting.

107. Zoological parks, botanical gardens, wildlife safari parks.

108. Zoological parks.

109. 1987, Dudhwa National Park.

110. Cryopreservation

111. *In vitro*, tissue culture, seed banks.

112. New Delhi, it conserves recalcitrant seeds which cannot be stored at low mositure content.

113. In 1992

114. Rio de Janeiro

115. (i) Conservation of biodiversity.
(ii) Sustainable utilisation of benefits of biodiversity.

116. 2002, global, regional, local.

CHAPTER SIXTEEN

ENVIRONMENTAL ISSUES

Topic 1 : Air Pollution and Its Control Measures

1. The enormous growth in human population size has impacted our environment. (*Give reason*)

   ____

2. Give a biological term for the following statement.
   Any undesirable change in physical, chemical or biological characteristics of air, land, water or soil. ____________________

3. Agents that bring about an undesirable change in natural resources are called as________________ (pollutants/waste products).

4. The Environment (Protection) Act was passed in ____________ .

5. Give the correct heading for the differences given below for types of pollutants.

A	B
These pollutants emerge directly from the source.	These pollutants are not directly emitted by the source, but are formed due to chemical reactions.
These pollutants are less toxic.	These pollutants are generally oxidising.
They persist in the form they are released as in the environment.	These pollutants are modified products.

   A.____________________ B. ____________________

6. Qualitative pollutant is a normal pollutant that has crossed their threshold value in the environment.
   (*Rewrite the statement by changing the underlined words.*)

   ____

7. A normal component of the environment becomes a quantitative pollutant when its concentration surpasses its threshold value.

   True ☐ False ☐

8. Most harmful environmental pollutants are ______________________ (biodegradable/non-biodegradable) wastes.

9. (i) ______________________ wastes are decomposed by bacteria. ______________________ wastes cannot be decomposed by bacteria.

(ii) Classify the following materials as biodegradable or non-biodegradable.

Sewage waste, DDT, Livestock waste, Cans, Broken glass, Paper

____

10. ______________________ is the presence of unwanted substances which alter the physical, biological and chemical properties of air.

11. How many of the following are primary air pollutants?

DDT, Nitrogen oxide, Ozone, Particulate matter, Sulphur dioxide, CO, Hydrocarbons, Ammonia, Toxic heavy metals, Olefins, Smog

____

12. Match the following columns.

Column I (Industrial sources of air pollution)	Column II (Emissions)
A. Automobiles	1. Lead, chromium
B. Plating	2. Particulates
C. Cement industry	3. Metal fumes
D. Metallurgy	4. Hydrocarbons

A.______________ B.______________ C.______________ D.______________

13. Consider the flowchart given below and answer the following questions that follows.
(i) Identify *A*, *B*, *C*, *D* and *E* in the flowchart given below.

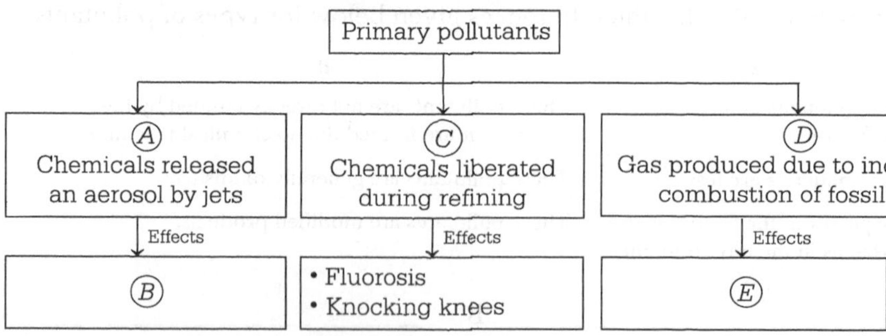

A.______________ B.______________ C.______________
D.______________ E.______________

(ii) Name the toxic compound formed by the reaction of *D* with haemoglobin in blood.

____

(iii) What does the above stated situation (mention in question II) lead to?

____

(iv) In animals, _____________ bring about abnormal calcification of bones and teeth.

(v) Which of the pollutant, i.e. *A, C* and *D* given in the flowchart represents the major cause of pollution in urban areas?_______________________

14. Thermal power stations produce minute suspended particles, constituting aerosols.

True ☐ False ☐

15. Give scientific term for the following.

The particulate matter formed by the combination of gas and liquid particles together.

____

16. A secondary pollutant is _______________ (carbon dioxide/smog).

17. Nitrogen oxides react with _______________ to produce a secondary pollutant _______________ .

18. In classical smog, the main components are secondary pollutants produced by the action of sunlight on emissions of automobiles and factories.

True ☐ False ☐

19. The harmful effects of the air pollution on organism depend on _______________ (size/concentration / duration of exposure) of pollutants.

20. Photochemical smog is formed at low temperature due to moving air.

True ☐ False ☐

21. Levels of air pollution can be estimated by the presence or absence of organisms called lichens.

Air pollution levels	Most common types of lichen present
Low	Shrubby
Medium	Leafy
High	Crusty

Given below is a study performed by environmental scientists carried out on lichen species at three different sites and obtained the results shown in the table below.

Number of Lichen Species Present

Site	Shrubby	Leafy	Crusty
A	0	5	19
B	3	2	0
C	7	14	2

(i) Which site according to you has the highest levels of air pollution? _______________

(ii) Give reason for your answer in question no. 21 (i). _______________

(iii) Lichens are sensitive to the presence of _______________ (sulphur dioxide/ dinitrogen) in air.

22. Who coined the term acid rain?

____

23. Acid rain is known to contain some acids. Name these acids._______________

24. From where do the acids get added in the rain?

____

25. Which gas leaked from Union Carbide's Pesticide Plant in December 1984 is responsible for Bhopal Gas Tragedy? (Methyl salicylate /Methyl isocyanate). _______________

26. In vehicles, catalytic converters are used to increase the efficiency of lead mixed petrol. (*Rewrite the statement by changing the underlined words.*)

____

27. Match the following columns.

	Column I (Pollutant removing devices)		Column II (Types of the pollutant removed)
A.	Electrostatic precipitator	1.	Removes SO_2
B.	Scrubber	2.	Removes particulate matter
C.	Catalytic converter	3.	Reduced emissions of poisonous gases
D.	Cyclonic separators	4.	Removes particulate matter using centrifugal force

A._______________ B._______________ C. _______________ D. _______________

28. Observe the diagram given below and answer the following questions.

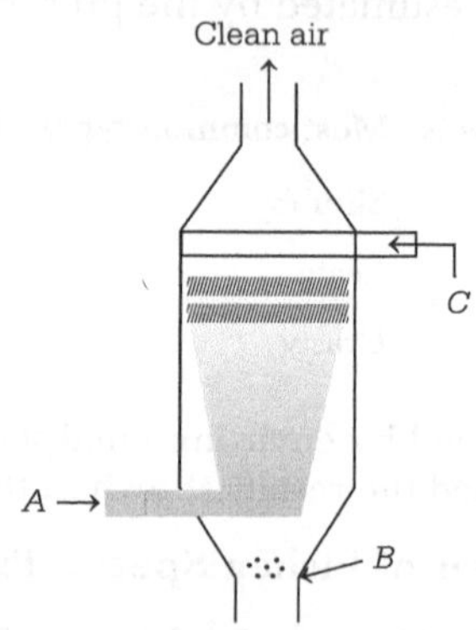

(i) The diagram represents a _______________ (scrubber/ cyclone separator).

(ii) Label the part from where the particulate matter rich waste is passed through _______________ (*A* /*B* /*C*).

(iii) In this device, the exhaust is passed through a spray of weak acid.

True ☐ False ☐

29. (i) Complete the labels *A*, *B* and *C* for a device used to control a specific type of air pollutant.

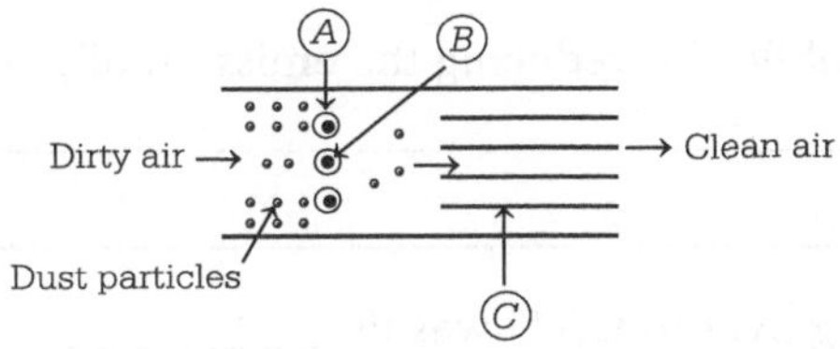

A. ____________

B. ____________

C. ____________

(ii) Name the device and the type of pollutant it helps in controlling.

____

30. Electrostatic precipitator has electrode wires that are maintained at several thousand volts, which produce a ____________ that releases electrons. These electrons attach to dust particles giving them a net ____________ . The collecting plates attract the charged dust particles.

31. Particulate matters of size 2.5 micrometres or less can directly enter our circulatory system. (*Rewrite the statement by changing the underlined words.*)

____

32. Expand CPCB. ____________

33. Justify the statement.
Particulate matters are responsible for causing the greatest harm to human health.

____

34. In catalytic converters, following reactions take place as the exhaust passes through it,
____________ ⟶ Carbon dioxide + ____________ . Identify these reactants.

35. Automobiles are a major cause for atmospheric pollution in rural cities.

True ☐ False ☐

36. How can we reduce the pollutants that are emitted by a large number of automobiles on the road?

____

37. Pick the odd one out with respect to metals used in catalytic converter.

Platinum, Silver, Palladium, Rhodium

____

38. Motor vehicles equipped with catalytic converters should use methanol.
(*Rewrite the statement by changing the underlined word.*)

____

39. Unleaded petrol is recommended for automobiles equipped with catalytic converters.
(*Give reason*) ____________

40. According to the Central Pollution Control Board (CPCB), particles that are responsible for causing great harm to human health are of ______________ diameter.

41. Name the device fitted to automobiles for reducing the emission of poisonous gases like NO_2 and CO. ______________

42. Why is CNG better than diesel? ______________

43. The main problem with switching over to CNG was its ______________ (uninterrupted supply/ being siphoned off).

44. What does mass emission standards signify?

____

45. Euro III norms stipulate that <u>nitrogen oxide</u> be controlled at 350 parts-per-million (ppm) in diesel and 150 ppm in petrol.
(*Rewrite the statement by changing the underlined word.*)

____

46. As per Euro III norms, ______________ should be contained at 42% of the concerned fuel.

47. The Bharat Stage II is equivalent to ______________ (Euro II/ Euro VI) norms, are no more applicable in any cities of India.

48. Complete the following table showing mass emission standards in India.

Types of vehicles	Norms	Cities of implementation
4 wheelers	Bharat stage IV	Throughout the country since (A)
(B)	Bharat stage IV	Throughout the country since 1st April 2017
2 wheelers	(C)	Throughout the country since (D)

A. ______________ B. ______________ C. ______________ D. ______________

49. In view of the severe pollution issue, Government of India has decided to shift to Bharat Stage VI emission standard for various category vehicles by ______________

50. Identify the system of our bodies that are deleteriously affected by air pollution from those mentioned in brackets (respiratory/digestive).

____

51. The Air (Prevention and Control of Pollution) Act came into force in 1981, but was amended in 1987. (*Give reason*)

____

52. Undesired levels of sound lead to air pollution.

True ☐ False ☐

53. Give any two causes of noise pollution.
(i) ______________ (ii) ______________

54. Maximum noise levels permissible during daytime in residential areas is ________ (60 dB/ 80 dB).

55. An exposure to extremely high sound level, 150 dB or more can impact us permanently. Justify. ______________

56. Sleeplessness, increased heartbeat, hearing disability, altered breathing pattern. What is common to these symptoms? ______________

57. How can we reduce noise levels around hospitals and schools?
____

58. A noise absorbent materials used in building to block noise is ______________ (styrofoam/cement).

59. Name the technique to reduce the noise pollution by planting four to six rows of green trees near the populated areas or along the roadside, industrial areas, etc.
____

Topic 2 : Water Pollution and Its Control

60. The degradation of water quality due to addition of undesirable materials. (*Give answer in one word.*)
____

61. Which act was passed to safeguard our water resources?
____

62. A non-point source of water pollution is ______________ (industrial effluents/agricultural runoffs).

63. Complete the table given below.

Industrial pollutants	Diseases
Arsenic	(A)
(B)	Itai itai
Nitrate	(C)
(D)	Minamata disease
Fluoride	(E)

A. ______________ B. ______________
C. ______________ D. ______________
E. ______________

64. Pick the odd one out.

Sewage, Oil spills, Leaching of soil, Coolants

____

65. A mere _______________ (0.1 /0.01)% impurities make domestic sewage unfit for human use.

66. Complete the following analogy with respect to organisms involved with pollution.
Lichens : Air pollution : : ___________________________ : Water pollution by sewage

67. Match the following columns.

	Column I (Water pollutant types)		Column II (Examples of pollutants)
A.	Suspended solids	1.	Faecal matter, cloth fibres, bacteria
B.	Colloidal materials	2.	Nutrients (phosphates, nitrates, etc.)
C.	Dissolved materials	3.	Sand, slit or clay

A.________________ B.________________ C. ________________

68. The diagram below shows the effect of polluting a river with untreated sewage water. What does graph *X* represent?

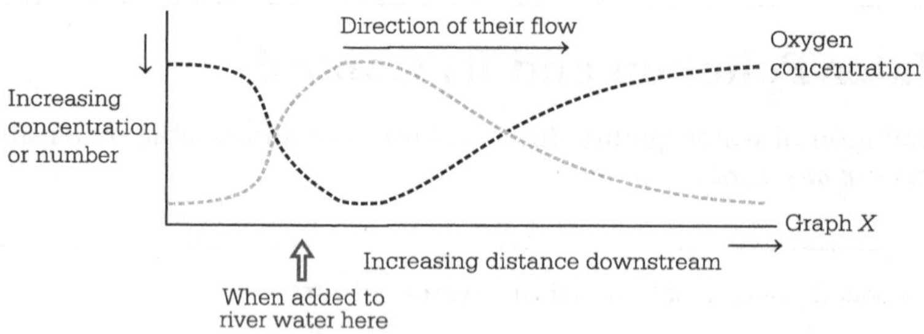

____

69. Dissolved material can be physically removed from the sewage.

True ☐ False ☐

70. Give a scientific term for the following sentence.

An measure of extent of pollution in a water body.______________________________

71. What is the relationship between BOD, microorganisms and the amount of biodegradable matter?

____

72. Complete the following analogy by expanding abbreviated words.
BOD : Biological Oxygen Demand :: COD : ___________________________

73. The overuse of fertilisers impact the aquatic animals by decreasing ______________ level and increasing ______________________ growth.

74. Name the scientific term used for the rapid growth of algae in the freshwater and marine water system.

75. Terror of Bengal is ______________________________ (*Eichhornia crassipes*/*Parthenium*).

76. The beautiful mauve-coloured flowers found on very appealingly-shaped floating plants in water bodies were introduced into India for their lovely flowers, but instead have caused havoc on our environment. (*Give reason*)

____

77. Untreated sewage should not be discharged into water bodies. (*Give reason*)

____

78. What is common to these diseases?

Dysentery, Typhoid, Jaundice, Cholera

____

79. Study the diagram given below and answer the following questions.

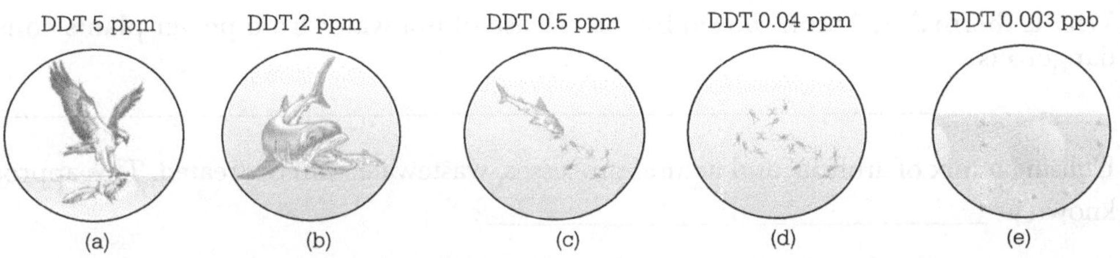

(i) Name the process taking place here. ____________________

(ii) The phenomenon occurring in the figure implies that ________________
(toxic materials are magnified/food is magnified) with each trophic level.

(iii) Given below are the organisms participating in a food chain.

Zooplankton, Large fish, Water, Fish eating birds, Small fish.

Place these organisms within the above figure in the correct sequence.

____

80. DDT residues pass rapidly through a food chain causing biomagnification because it is water soluble.

(*Rewrite the statement by changing the underlined words.*)

____

81. High concentrations of DDT have caused a decline in bird populations. (*Give reason*)

____

82. Give the scientific term for the following statement.

The natural ageing of water bodies. ____________________

83. Which of the following have faster growth rates - natural eutrophication or cultural eutrophication ? (*Give reason*)

____

84. What contributes to the natural ageing of a water body?

____

85. As the fertility of the water body increases, _______________ begin to deposit on the bottom of water bodies.

86. The first plants to grow in the shallows are _______________ (marsh/floating) plants.

87. Nitrogen and phosphorus, carbon and sulphur. Which of the following combinations encourage the acceleration of eutrophication ?

____

88. How many of the following are associated with a lake undergoing accelerated eutrophication?

Growth of algae, Unsightly scum, White foam, Unpleasant odors, Increased number of fishes

____

89. Why is thermal pollution caused by the release of hot water from power plants considered dangerous?

____

90. Utilising a mix of artificial and natural processes, wastewater can be treated. This approach is known as _______________.

91. Given below is a flowchart showing the stages of integrated treatment of wastewater.

Conventional sedimentation
↓
(A)
↓
(B)
↓
Biological treatment

(i) Identify labels *A* and *B*. A. _______________ B. _______________

(ii) After the cleaning of water by natural processes, toxic _______________ (heavy metals/ plastic), remains in the effluent.

(iii) In this innovative approach, the biologists developed a series of six connected marshes over 60 hectares of marshland. Appropriate plants, algae, fungi and bacteria were seeded into this area for what purpose? _______________

(iv) An example where this approach has been established is

____

(v) Biological treatment of wastewater aims to _______________ (reduce BOD/ increase BOD).

92. What is FOAM?

____

93. Is there a possibility of water not being necessary to dispose off human waste?

____

94. Ecological sanitation is a sustainable system for handling human excreta by using toilets which have connection to sewage treatment plants.
(*Rewrite the statement by changing the underlined words.*)

____

95. The dry composting toilets increase the reliance on chemical fertilisers.
True ☐ False ☐

96. EcoSan toilets are working in parts ____________ and ____________ (Kerala/Maharashtra/Sri Lanka).

Topic 3 : Solid Wastes, Agrochemicals and Radioactive Wastes

97. Give one word for the following statement.
The discarded or left over waste that goes in trash.

____

98. Pick the odd one out.

Paper, Food waste, Fly ash, Textile

____

99. Write the names of the category of solid wastes against the waste materials mentioned below.

(i) Slack and slag, rock trailings, mine dusts. ____________

(ii) Blood pouches, used needles, disposable surgical instruments, expired medicines .

____

(iii) Irreparable phones, computers and other electronic goods. ____________

(iv) Wastes generated from cancer treatment centres, nuclear plants. ____________

(v) Waste accumulating during growing and harvesting seasons. ____________

100. An idea behind solid waste management is to ____________ (stop waste generation/control of waste generation).

101. Bio-medical waste can be effectively managed by the thermal process.
True ☐ False ☐

102. The simplest and most common method used in the cities is to collect and dump the waste. (*Give answer in one word.*)

____

103. Given below are some methods for waste disposal. How many of them are used for solid waste disposal methods?

Land filling, Natural breakdown, Recycling, Open direct burning, Sewage treatment system

____

104. In a sanitary landfill, wastes are dumped in a _______________ after ____________, and covered with dirt.

105. There is a risk of _________________________ from the sites of sanitary landfills.

106. As the matter inside the sanitary landfill breaks down, it generates different gases including ___________________ (nitrogen/methane).

107. Burning reduces the volume of the wastes yet it is not a preferred method of waste disposal. (*Give reason*)

____

108. What are the 3Rs for effective waste management?

____

109. All wastes that we generate can be categorised into three types, i.e. biodegradable, recyclable and the non-biodegradable. How does this step help in waste management and disposal?

____

110. To go eco-friendly, we can _______________________________ (carry cloth or other natural fibre carry-bags/ ask for plastic bags).

111. Polyblend is prepared from _______________________________________.

112. Polyblend and bitumen when blended and used together increased the life of roads by 3X. (*Give reason*) _______________________________

113. Complete the following table given below.

Solid wastes	Methods of disposal	Descriptions
Tin cans, metal wastes, plastic bags, paper	(A)	Rag pickers collect wastes
(B)	Composting	Decomposition in pits
Non-recyclable wastes	(C)	Anaerobic combustion at 165°C
Hospital wastes	Incineration	(D)
Oil spills	(E)	Microorganisms remove or neutralise pollutants from a contaminated site
(F)	Landfills/ Incinerated	Buried in trenches or burned

A. _______________ B. _______________ C. _______________ D. _______________

E. _______________ F. _______________

114. The recycling in developing countries often involves specifically built facilities.

True ☐ False ☐

115. Pick the odd one out with respect to chemicals.

Pesticides, Herbicides, Manure, Fungicides

____

116. What is common to the following terms?

Algal blooms, Accelerated eutrophication, Biomagnification

____

117. Complete the following analogy with respect to polyblend.

Ahmed Khan : Polyblend :: Ramesh Chandra Dagar : ______________

118. Integrated organic farming includes how many of the components mentioned below?

Beekeeping, Fisheries, Dairy management, Water harvesting, Composting, Agriculture, Hybridisation

____

119. Farming is an extremely economical and sustainable venture. (*Give one word*)

____

120. Complete the following passage.

In integrated farming, cattle excreta can be used as ______________. Crop waste is used to create ______________, which can be used as a ______________ or can be used to generate ______________ for satisfying the energy needs of the farm.

121. Nuclear energy is a polluting way for generating electricity. (*Give reason*)

____

122. Three Mile Island and Chernobyl disaster are associated with water pollution. (*Rewrite the statement by changing the underlined word.*)

____

123. Radiations given off by nuclear waste is extremely damaging to living organisms because they create ______________

124. Alpha, beta and gamma, among ______________ is least damaging and ______________ is most lethal.

125. Exposure to radiations in low amounts causes ______________ (respiratory disease/cancer).

126. Nuclear waste is an extremely potent pollutant and has to be dealt with utmost caution. List any one precaution associated with disposal of such pollutants.

____

127. Radioactive wastes are buried about 10 km deep below the earth's surface. (*Rewrite the statement by changing the underlined word.*)

____

Topic 4 : Greenhouse Effects, Global Warming and Ozone Depletion

128. Give a biological term for following statement.

A naturally occurring phenomenon responsible for heating of earth's surface and atmosphere.

____

129. Observe the diagram given below and identify the relative contributions of different greenhouse gases.

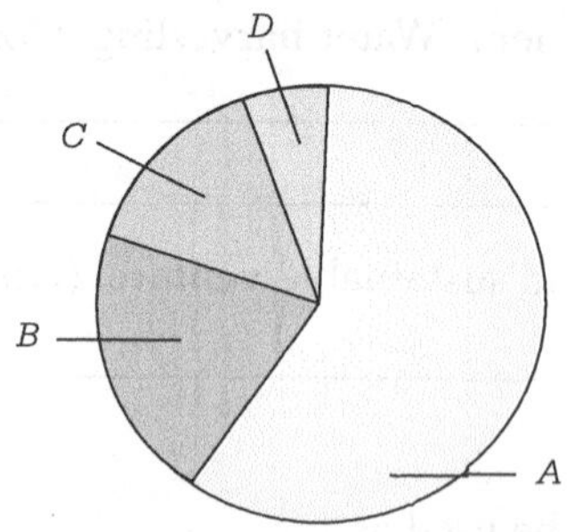

A. ______________

B. ______________

C. ______________

D. ______________

130. If the greenhouse phenomenon was absent, the average temperature of earth would be

____

131. The greenhouse gases are so named because (*Answer in one line*)

____

132. Clouds and gases absorb about one-fourth of the incoming solar radiation and reflects some of it.
(*Rewrite the statement by changing the underlined words.*)

____

133. Greenhouse gases absorb long waves radiation from the earth, emit it again towards the earth. This cycle continues to repeat. What would happen if the greenhouse gases responsible for this phenomenon increases?

____

134. El Nino effect is a consequence of global warming.

True ☐ False ☐

135. The El Nino effect represents the disturbed rainfall pattern in some areas .
(*Rewrite the statement by changing the underlined words.*)

____

136. Give any two measures we can use to control global warming.

____

137. Complete the following analogy with respect to ozone layers.

Good ozone : ______________ : : Bad ozone : Troposphere

____

138. What is the function of the ozone layer?

____

139. Use figure given below and answer the following questions.

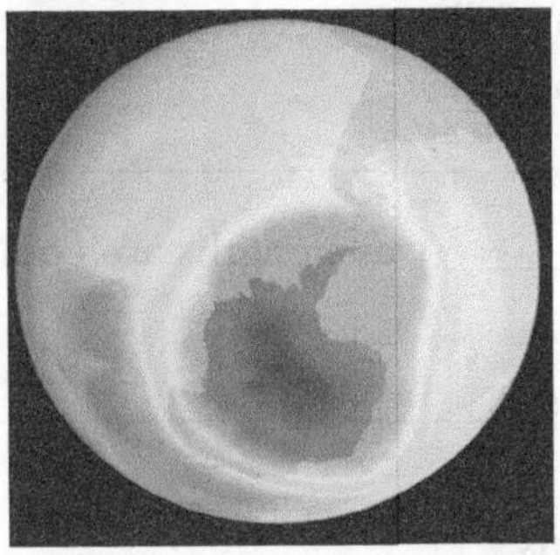

(i) The figure represents ____________ (ozone layer in stratosphere/ozone hole).

(ii) How do we measure ozone thickness to determine the depletion of the ozone layer?

____

140. Ozone hole was first observed over Asian continent.

True ☐ False ☐

141. Justify. UV rays are highly injurious to living organisms.

____

142. Protection against UV radiation is proportional to ____________

143. Complete the following reaction.

O_2 (Molecular oxygen) $\xrightarrow{A}$ O_3 (Ozone), identify *A* and also indicate the layer of atmosphere it occurs. ______________________________

144. When ozone depletion occurs? (*Answer in one line*)

____

145. Complete the following analogy.

Greenhouse gases : Global warming :: ____________ : Ozone depletion

146. Jet propellants, refrigerants, solid plastic foams, what is common to these substances?

____

147. Complete the following reaction.

$$\text{CFCs} \xrightarrow{\text{UV rays}} (A) + O_3 \longrightarrow O_2 + (B)$$

A. ______________________________

B. ______________________________

148. Chlorine atoms get exhausted during ozone depletion.

True ☐ False ☐

149. CFCs are added to the stratosphere, they have temporary effects on ozone levels. (*Rewrite the statement by changing the underlined words.*)

____

150. Give a scientific term for the following sentence.
A large area of thinned ozone layer.

____

151. UV radiations of __________ wavelengths are completely absorbed by the earth's atmosphere.

152. A high dose of UV-B causes ________________ (snow blindness/sunburns) in the cornea of humans.

153. Match the following columns.

Column I	Column II
A. Kyoto protocol	1. Climate change
B. Montreal protocol	2. Protection of ozone layer
C. Vienna convention	3. Reduce the emission of greenhouse gases
D. Earth summit	4. Control emission of ozone depleting substances

A.________ B.________ C. ________ D. ________

Topic 5 : Improper Resource Utilisation and Deforestation

154. Overcultivation, unrestricted grazing, deforestation, poor irrigation - the most likely consequence of these activities ________________ (soil erosion and desertification/waterlogging).

155. Irrigation without proper drainage of water leads to ________________.

156. Waterlogging draws salt under the surface of the soil.

True ☐ False ☐

157. Increased salt content is inimical to the growth of crops and is extremely damaging to agriculture. (*Give reason*) ________________________

158. Complete the following analogy with respect to forest as natural resources.
________________ : Cutting of trees :: Reforestation : Planting of trees

159. Almost 60% forests have been lost in the tropics due to indiscriminate cutting of trees. (*Rewrite the statement by changing the underlined word.*)

____

160. The forest cover shrunk to __________% at the end of century.

161. The National Forest Policy (1988) of India has recommended ____________ % forest cover for the plaine and __________ % for the hills.

162. A cultivation practice responsible for deforestation is ____________________ (cyclic agriculture/Jhum cultivation).

163. Complete the passage given below.

In ____________________ agriculture, the farmers cut down the trees of the forest and________________ the plant remains. The ash is used as a ________________ and the land is then used for farming or cattle grazing. After cultivation, the area is left for several years so as to allow its recovery.

164. What is the difference between Jhum cultivation of the past and now-a-days?

____

165. An outcome of deforestation is ____________________ (enhanced carbon dioxide concentration/ soil salinity) in the atmosphere.

166. Deforestation leads to loss of biodiversity due to ____________________

167. How many of the following are consequences of deforestation?

Loss of biodiversity, Waterlogging, Cataracts, Disturbed hydrologic cycle, Soil erosion, Desertification.

____

168. Deforestation can be reversed using ____________________

169. Name the award given to individuals or communities from rural areas who have shown extraordinary courage and dedication in protecting wildlife.

____

170. Amrita Devi and Sundar Lal Bahuguna are names associated with which movement environment protection?

____

171. Chipko movement was organised for the protection of water bodies .
(*Rewrite the statement by changing the underlined word.*)

____

172. Give the full form of JFM. ____________________

173. When did JFM came into force? ____________________

174. What was the agenda behind JFM? (*Answer in one line*)

____

Answers

1. Human population creates tremendous pressure on our natural resources and is also contributing to pollution of natural resources.
2. Pollution
3. Pollutants
4. 1986
5. Primary pollutants, secondary pollutants
6. Harmful substance added by human activities to
7. True
8. Non-biodegradable
9. (i) Biodegradable non-biodegradable
   (ii) Sewage waste, livestock waste and paper are biodegradable wastes, while DDT, cans, broken glass are non-biodegradable wastes.
10. Air pollution
11. 8 (DDT, nitrogen oxide, particulate matter, sulphur dioxide, CO, hydrocarbons, ammonia, toxic heavy metals).
12. A-4, B-3, C-2, D-1
13. (i) *A* - Chlorofluorocarbons, *B* - Ozone depletion, *C* - Fluorides and silica, *D* - Carbon monoxide, *E* - Impaired oxygen transport in amimals
    (ii) Carboxyhaemoglobin
    (iii) It leads to very low oxygen conditions in body.
    (iv) Fluorides
    (v) Option *D*, i.e. carbon monoxide released due to incomplete combustion of fuels.
14. False. Thermal power stations produce minute suspended particles, constituting the fly ash.
15. Aerosols
16. Smog
17. Hydrocarbons, PAN(Peroxyacyl Nitrates)
18. False. In classical smog, the main components are mostly primary pollutants.
    It is formed by the condensation of water vapour with H_2O and SO_2 over dust particles.
19. All of them
20. False. Photochemical smog is formed at high temperature due to still air.
21. (i) Site A
    (ii) It has the highest population of crusty lichens, which grows in the areas where air pollution level is very high.
    (iii) Sulphur dioxide
22. Robert Angus Smith in 1872
23. Acid rain contains H_2SO_4 (sulphuric acid), HNO_3 (nitric acid) and H_2CO_3 (carbonic acid).
24. HNO_3 is formed by the oxidation of nitric oxide (NO) present in the air, H_2CO_3 is derived from carbon dioxide, H_2SO_4 is formed by the oxidation of SO_2 released from automobiles.
25. Methyl isocyanate
26. Reduce the emission of poisonous gases.
27. A-2, B-1, C-3, D-4.
28. (i) Scrubber.
    (ii) *B*
    (iii) False. In this device, the exhaust is passed through a spray of lime or water (*C*).
29. (i) *A* – Discharge corona
    B – Negatively charged wire
    C – Collecting plate
    (ii) Electrostatic precipitator, it helps in controlling particulate matter.
30. Corona, negative charge
31. Be inhaled deep into the lungs.
32. Central Pollution Control Board
33. Particulate matter causes breathing and respiratory symptoms, irritation, inflammations and damage to the lungs and premature deaths.
34. The reactants are unburnt hydrocarbons, nitrogen gas
35. False. Automobiles are a major cause for atmospheric pollution in metro cities.
36. Proper maintenance of automobiles along with the use of lead free petrol or diesel can reduce the pollutants emitted by them.
37. Silver is the odd one out, because other metals are used as catalysts in catalytic converters.
38. Lead free petrol
39. This is because lead in the petrol inactivates the catalyst of catalytic converters.
40. 2.50 μm
41. Catalytic converters
42. CNG burns most efficiently, unlike petrol or diesel and is much cheaper than diesel and petrol.
43. Uninterrupted supply
44. Mass emission standards are legally set quantitative limits on the permissible levels of specific air pollutants released into the atmosphere.
45. Sulphur
46. Aromatic hydrocarbons
47. Euro II
48. *A* – April 2020, *B*–3 wheelers
    C – Bharat stage IV, *D*–April 2020
49. April 1, 2020
50. Respiratory system
51. Air (Prevention and Control of Pollution) Act came into force in 1981, but was amended in 1987 to include noise as an air pollutant.
52. False. Undesired levels of sound lead to noise pollution.
53. Causes of noise pollution are
    (i) Burning crackers,
    (ii) Loudspeakers
54. 80 dB

55. Noise pollution at 150 dB can permanently damage the hearing abilities of humans.
56. They are all discomforts caused by noise pollution.
57. Using sound absorbent in buildings to reduce noise, green mufflers zone near schools, hospitals also reduce noise.
58. Styrofoam
59. Green mufflers
60. Water pollution
61. Water (Prevention and Control of Pollution) Act, 1974.
62. Industrial effluents.
63. *A* – Black foot disease, *B* – Cadmium, *C* – Blue baby syndrome, *D* – Mercury, *E* – Fluorosis
64. Leaching of soil is the odd one out, as it is a natural source of water pollution, while others are anthropogenic sources.
65. 0.1
66. Coliforms (*E. coli*)
67. A-3, B-1, C-2
68. Graph *X* represents the Biological Oxygen Demand (BOD) of water body.
69. False. Dissolved materials are nutrients such as phosphates, nitrates and organic compounds, etc., which are difficult to remove physically.
70. Biological Oxygen Demand (BOD)
71. Higher the BOD, higher will be the number of microorganisms (bacteria) and higher will be the rate of conversion of biodegradable organic components present in the sewage.
72. Chemical Oxygen Demand
73. Oxygen, algal
74. Algal bloom
75. *Eichhornia crassipes*
76. They grow at a very fast rate and block the surface of water bodies, cutting off both the sunlight and oxygen supply essential for aquatic life.
77. Untreated sewage may contain many undesirable pathogenic microorganisms and its disposal into water may cause an outbreak of serious diseases.
78. These all diseases are water borne, i.e. due to consumption of polluted water.
79. (i) Biomagnification
    (ii) Toxic materials are magnified.
    (iii) (a) Fish eating birds, (b) Large fish, (c) Small fish, (d) Zooplankton, (e) Water.
80. Liposoluble
81. High concentrations of DDT disturb calcium metabolism in birds, causing thinning of egg shells and their premature breaking, leading to a decline in bird populations.
82. Eutrophication
83. Cultural or accelerated eutrophication is much faster because of the presence of excessive nutrients from discharged sewage and agricultural runoffs.
84. Biological enrichment of water
85. Organic pollutants
86. Marsh
87. Nitrogen and phosphorus encourage the acceleration of eutrophication processes.
88. 3 (Growth of algae, unsightly scum and unpleasant odor).
89. It decreases the solubility of oxygen and it changes the breeding cycles of various aquatic organisms.
90. Integrated wastewater treatment
91. (i) *A* - Filtering, *B* - Chlorine treatment
    (ii) Heavy metals
    (iii) To neutralise, absorb and assimilate the pollutants. Hence, as the water flows through the marshes, it gets purified naturally.
    (iv) Arcata, situated along the Northern coast of California.
    (v) Reduce the BOD
92. FOAM is Friends of the Arcata Marsh who are responsible for the upkeep and safeguarding of integrated wastewater treatment project.
93. Yes, by ecological sanitation.
94. Using dry composting toilets
95. False. The dry composting toilets reduce the reliance on chemical fertilisers.
96. Kerala, Sri Lanka
97. Solid wastes
98. Fly ash is the odd one out because it is an industrial waste, while others are municipal waste.
99. (i) Mining wastes
    (ii) Biomedical wastes
    (iii) *e*-wastes
    (iv) Radioactive wastes
    (v) Agricultural wastes
100. Control of waste generation
101. True
102. Landfills
103. 4 (Land filling, natural breakdown, recycling, open direct burning). Sewage treatment is used for polluted water.
104. Depression or trench, compaction
105. Chemical leakage
106. Methane
107. This is because waste is generally not burnt to completion, it produces odurs and air pollutants and open dumps often serve as the breeding ground for rats and flies.
108. 3Rs are reduce, reuse and recycle.
109. Segregation of waste helps in identifying the best method for their disposal and helps in reducing the amount of waste generated.
110. Carry cloth or other natural fibre carry-bags.
111. Recycled modified plastic
112. Addition of polyblend, enhanced the bitumen's water repellant properties and helped in

increasing the road life by a factor of three.

113. *A* - Recycle, *B* - Organic matter, *C* - Pyrolysis, *D* - Aerobic combustion at 900-130°C, *E* - Bioremediation, *F* - *e*-wastes.

114. False. The recycling in developing countries often involves manual participation.

115. Manure is the odd one out because it is a component of organic farming, while others are chemical fertilisers.

116. All of them are the result of increasing amount of chemical fertilisers.

117. Organic farming

118. 5(Beekeeping, dairy management, water harvesting, composting and agriculture).

119. Integrated organic farming

120. Manure, compost, natural fertiliser, natural gas

121. The use of nuclear energy for generating electricity has two serious inherent problems, i.e. accidental leakage and disposal of radioactive wastes.

122. Radioactive

123. Mutations

124. Alpha radiation, gamma radiation

125. Cancer

126. Nuclear waste should be pretreated before disposal to remove any remaining radioactivity and stored in shielded containers.

127. 500 m

128. Greenhouse effect

129. The contribution of various greenhouse gases are

A–CO_2- 60%, *B*–CH_4- 20%, *C*–CFCs - 14%, *D*–N_2O -6%

130. – 18°C

131. They trap the heat radiated from the earth's surface.

132. Gases reflect about one-fourth of the incoming solar radiation and absorb.

133. Increase in the level of greenhouse gases will lead to considerable heating of earth causing global warming.

134. True

135. Odd climatic changes taking place in some areas on the earth.

136. Improving the efficiency of energy usage, reducing deforestation.

137. Stratosphere

138. It acts as a shield absorbing ultraviolet radiation from the sun.

139. (i) Ozone hole

(ii) Ozone is measured in Dobson's unit.

140. False. Ozone hole was first observed over Antarctica.

141. UV rays are harmful because DNA and proteins of living organisms preferentially absorb UV rays and its high energy breaks the chemical bonds within these molecules.

142. Thickness of the ozone layer

143. *A* is UV rays, stratosphere

144. Ozone depletion occurs when the balance between production and degradation of ozone in the stratosphere is disturbed.

145. Chlorofluorocarbons

146. They all release CFCs in the atmosphere.

147. *A* – Cl, *B*–ClO

148. False. Cl atoms are not consumed during ozone depletion.

149. Permanent and continuing effects

150. Ozone hole

151. Shorter

152. Snow blindness

153. A-3, B-4, C-2, D-1

154. Soil erosion and desertification.

155. Waterlogging

156. False. Waterlogging draws salt to the surface of the soil.

157. Increased salt makes the soil toxic, cause death of roots, high osmotic stress and nutritional disorders.

158. Deforestation

159. 40%

160. 21.54

161. 33, 67

162. Jhum cultivation

163. Slash and burn, burn, fertiliser

164. The time gap for recovery of cultivated land has decreased.

165. Enhanced carbon dioxide concentration

166. Habitat destruction

167. 4 (Loss of biodiversity, disturbed hydrologic cycle, soil erosion and desertification).

168. Reforestation

169. Amrita Devi Bishnoi Wildlife Protection Award

170. Chipko movement

171. Forests

172. Joint Forest Movement

173. 1980

174. The agende behind JFM was to work closely with the local communities for protecting and managing forests.